A117.1-2003

American National Standard

Accessible and Usable Buildings and Facilities

Secretariat

**International Code Council
Chicago District Office
4051 W. Flossmoor Road
Country Club Hills, IL 60478-5795**

Approved November 26, 2003

**American National Standards Institute, Inc.
25 West 43rd Street — Fourth Floor
New York, NY 10036**

First Printing: May 2004
Second Printing: August 2006

First Published: May 2004

ISBN 1-58001-102-0

COPYRIGHT © 2004

By

INTERNATIONAL CODE COUNCIL, INC.

PRINTED IN THE U.S.A.

American National Standard

Approval of an American National Standard requires verification by ANSI that the requirements for due process, consensus, and other criteria for approval have been met by the standards developer.

Consensus is established when in the judgement of the ANSI Board of Standards Review, substantial agreement has been reached by directly and materially affected interests. Substantial agreement means much more than a simple majority, but not necessarily unanimity. Consensus requires that all views and objections be considered, and that a concerted effort be made toward their resolution.

The use of American National Standards is completely voluntary; their existence does not in any respect preclude anyone, whether he or she has approved the standards or not, from manufacturing, marketing, purchasing, or using products, processes, or procedures not conforming to the standards.

The American National Standards Institute does not develop standards and will in no circumstances give an interpretation of any American National Standard. Moreover, no person shall have the right or authority to issue an interpretation of an American National Standard in the name of the American National Standards Institute. Requests for interpretations should be addressed to the secretariat or sponsor whose name appears on the title page of this standard.

CAUTION NOTICE: This American National Standard may be revised or withdrawn at any time. The procedures of the American National Standards Institute require that action be taken periodically to reaffirm, revise, or withdraw this standard. Purchasers of American National Standards may receive current information on all standards by calling or writing the American National Standards Institute.

FOREWORD

(The information contained in this foreword is not part of this American National Standard (ANS) and has not been processed in accordance with ANSI's requirements for an ANS. As such, this foreword may contain material that has not been subjected to public review or a consensus process. In addition, it does not contain requirements necessary for conformance to the standard.)

Development

The 1961 edition of ANSI Standard A117.1 presented the first criteria for accessibility to be approved as an American National Standard and was the result of research conducted by the University of Illinois under a grant from the Easter Seal Research Foundation. The National Easter Seal Society and the President's Committee on Employment of People with Disabilities became members of the Secretariat, and the 1961 edition was reaffirmed in 1971.

In 1974, the U.S. Department of Housing and Urban Development joined the Secretariat and sponsored needed research, which resulted in the 1980 edition. After further revision that included a special effort to remove application criteria (scoping requirements), the 1986 edition was published and, when requested in 1987, the Council of American Building Officials (CABO) assumed the Secretariat. Central to the intent of the change in the Secretariat was the development of a standard that, when adopted as part of a building code, would be compatible with the building code and its enforcement. The 1998 edition largely achieved that goal. In 1998, CABO became the International Code Council (ICC).

2003 Edition

New to the 2003 edition are criteria for elements and fixtures primarily for children's use; enhanced reach range criteria; transportation facilities; additional provisions for assembly areas; and an addition and rearrangement for accessible dwelling and sleeping units. These new criteria are intended to provide a level of coordination between the accessible provisions of this standard and the Fair Housing Accessibility Guidelines (FHAG) and the Americans with Disabilities Act Accessibility Guidelines (ADAAG). Illustrative figures are numbered the same as corresponding text to simplify the use of the Standard. Unless specified otherwise, figures are not part of the Standard. Should a figure appear to illustrate criteria that differ with the text of the Standard, the criteria stated in the text govern.

ANSI Approval

This Standard was processed and approved for submittal to ANSI by the Accredited Standards Committee A117 on Architectural Features and Site Design of Public Buildings and Residential Structures for Persons with Disabilities. ANSI approved the 2003 edition on November 26, 2003. Committee approval of the Standard does not necessarily imply that all Committee members voted for its approval.

Adoption

ICC/ANSI A117.1-2003 is available for adoption and use by jurisdictions internationally. Its use within a governmental jurisdiction is intended to be accomplished through adoption by reference in accordance with proceedings establishing the jurisdiction's laws.

Formal Interpretations

Requests for Formal Interpretations on the provisions of ICC/ANSI A117.1-2003 should be addressed to: ICC, Chicago District Office, 4051 W. Flossmoor Road, Country Club Hills, IL 60478-5795.

Maintenance—Submittal of Proposals

All ICC standards are revised as required by ANSI. Proposals for revising this edition are welcome. Please visit the ICC website at www.iccsafe.org for the official "Call for proposals" announcement. A proposal form and instructions can also be downloaded from www.iccsafe.org.

ICC, its members and those participating in the development of ICC/ANSI A117.1-2003 do not accept any liability resulting from compliance or noncompliance with the provisions of ICC 300-2002. ICC does not have the power or authority to police or enforce compliance with the contents of this standard. Only the governmental body that enacts this standard into law has such authority.

Accredited Standards Committee A117 on Architectural Features and Site Design of Public Buildings and Residential Structures for Persons with Disabilities

At the time of ANSI approval, the A117.1 Committee consisted of the following members:

Chair . . . . Kenneth M. Schoonover, PE
Vice Chair . . . . Edward A. Donoghue, CPCA
A117 Committee Secretary . . . . Lawrence Brown, CBO

Organizational Member	Representative
Accessibility Equipment Manufacturers Association (AEMA) (**PD**)	Kevin Brinkman
	Gregory L. Harmon (Alt)
American Bankers Association (ABA) (**BO**)	Nessa Feddis
American Council of the Blind (ACB) (**CU**)	Patricia Beattie
	Krista Merritt (Alt)
American Hotel and Lodging Association (AHLA) (**BO**)	Kevin Maher
	Jerry Gross (Alt)
American Institute of Architects (AIA) (**P**)	Mark W. Wales, Associate AIA, CBO
	Larry M. Schneider (Alt)
American Occupational Therapy Association (AOTA) (**P**)	S. Shoshana Shamberg
American Society of Interior Designers (ASID) (**P**) . . .	Samantha McAskill, ASID
	Barbara J. Huelat, ASID, IIDA (Alt)
American Society of Plumbing Engineers (ASPE) (**P**)	Robert H. Evans, Jr., CIPE/CPD
	Julius A. Ballanco, P.E. (Alt)
American Society of Safety Engineers (ASSE) (**P**) . . .	Dr. William Marletta, PhD, CSP
	John B. Schroering, P.E., C.S.P. (Alt)
American Society of Theatre Consultants (ASTC) (**P**) .	William Conner, ASTC
	R. Duane Wilson, ASTC (Alt)
Association for Education & Rehabilitation of the Blind & Visually Impaired (AERBVI) (**P**)	Billie Louise "Beezy" Bentzen, PhD
Builders Hardware Manufacturers Association, Inc (BHMA) (**PD**)	Michael Tierney
	Richard Hudnut (Alt)
Building Owners and Managers Association International (BOMA) (**BO**)	Lawrence G. Perry, AIA
	Ron Burton (Alt)
Disability Rights Education and Defense Fund (DREDF) (**CU**)	Marilyn Golden
	Logan Hopper (Alt)
International Association of Amusement Parks and Attractions (IAAPA) (**BO**)	John Paul Scott, AIA, NCARB
International Code Council (ICC) (**R**)	Kimberly Paarlberg, AIA
	Phil Hahn, AIA (Alt)
International Sign Association (ISA) (**PD**)	Teresa Cox
	James L. Chapman (Alt)
	Muhammad Kahn (2nd Alt)
Little People of America, Inc. (LPA) (**CU**)	Tricia Mason
	C. Angela Van Etten, J.D.
Montgomery County Department of Permitting Services (MCDPS) (**R**)	Shahriar Amiri, CBO
	Thomas Heiderer (Alt)
National Apartment Association (NAA) (**BO**)	Ronald G. Nickson
National Association of Home Builders (NAHB) (**BO**) .	Jeffrey T. Inks
	Richard A. Morris (Alt)
National Association of the Deaf (NAD) (**CU**)	Don Sievers
	Nancy J. Bloch (Alt)

National Conference of States on Building Codes
and Standards (NCSBCS) (**R**) . . . . Curt Wiehle

National Electrical Manufacturers Association
(NEMA) (**PD**) . . . . Rein Haus
  Scott Edwards (Alt)
  James Shuster (2nd Alt)

National Elevator Industry, Inc (NEII) (**PD**) . . . . . . Edward A. Donoghue, CPCA
  George A. Kappenhagen (Alt)
  Barry Blackaby (2nd Alt)

National Fire Protection Association
(NFPA) (**R**) . . . . Allan B. Fraser
  Ron Coté, PE (Alt)

New Mexico Governor's Committee on Concerns of the
Handicapped (NMGCCH) (**CU**) . . . . Hope Reed
  Anthony H. Alarid (Alt)

Paralyzed Veterans of America (PVA) (**CU**) . . . . . Carol Peredo Lopez, AIA
  Mark H. Lichter, AIA (Alt)

Plumbing Manufacturers Institute (PMI) (**PD**) . . . . David Viola
  Barbara C. Higgens (Alt)

Self Help for Hard of Hearing People, Inc.
(SHHH) (**CU**) . . . . Brenda Battat
  Timothy P. Creagan (Alt)

Society for Environmental Graphic Design
(SEGD) (**P**) . . . . Kenneth A. Ethridge, Jr., AIA, **RIBA**
  Ann Makowski (Alt)

Stairway Manufacturers Association (SMA) (**PD**) . . . Tim Moss
  David Cooper (Alt)

U.S. Architectural & Transportation Barriers Compliance
(Access) Board (ATBCB) (**R**) . . . . Marsha K. Mazz
  Scott J. Windley (Alt)

U.S. Department of Agriculture (USDA) (**R**) . . . . . . Larry B. Fleming
  Samuel Hodges, III (Alt)

U.S. Department of Housing and Urban Development
(HUD) (**R**) . . . . Cheryl D. Kent
  Louis F. Borray (Alt)

United Cerebral Palsy Association, Inc. (UCPA) (**CU**) . Robert Dale Lynch, FAIA
  Gus Estrella (Alt)

United Spinal Association (**CU**) . . . . Brian D. Black
  Dominic Marinelli (Alt)

World Institute on Disability (WID) (**CU**) . . . . Hale Zukas

Individual Members

Todd Andersen (P)
George P. McAllister, Jr. (P)
Jake L. Pauls, CPE (P)
John P. S. Salmen, AIA (P)
Kenneth M. Schoonover, PE (P)

Acknowledgment

The updating of this standard over the past 5 years could only be accomplished by the hard work of not only the current committee members listed at the time of approval but also the many committee members who participated and contributed to the process over the course of development. ICC recognizes their contributions as well as those of the participants who, although not on the committee, provided valuable input during this update cycle.

INTEREST CATEGORIES

Builder/Owner/Operator (BO) - Members in this category include those in the private sector involved in the development, construction, ownership and operation of buildings or facilities; and their respective associations.

Consumer/User (CU) - Members in this category include those with disabilities, or others who require accessibility features in the built environment for access to buildings, facilities and sites; and their respective associations.

Producer/Distributor (PD) - Members in this category include those involved in manufacturing, distributing, or sales of products; and their respective associations.

Professional (P) - Members in this category include those qualified to engage in the development of the body of knowledge and policy relevant to their area of practice, such as research, testing, consulting, education, engineering or design; and their respective associations.

Regulatory (R) - Members in this category include federal agencies, representatives of regulatory agencies or organizations that promulgate or enforce codes or standards; and their respective associations.

Individual Expert (IE) (Nonvoting) - Members in this category are individual experts selected to assist the consensus body. Individual experts shall serve for a renewable term of one year and shall be subject to approval by vote of the consensus body. Individual experts shall have no vote.

Category	Number
Builder/Owner/Operator - **(BO)**	6
Consumer/User - **(CU)**	10
Professional - **(P)**	8
Producer/Distributor - **(PD)**	7
Regulatory - **(R)**	7
TOTAL	**38**

Contents

List of Figures

Chapter 1. Application and Administration

101 Purpose

The technical criteria in Chapters 3 through 9, and Sections 1002, 1003 and 1005 of this standard make sites, facilities, buildings and elements accessible to and usable by people with such physical disabilities as the inability to walk, difficulty walking, reliance on walking aids, blindness and visual impairment, deafness and hearing impairment, incoordination, reaching and manipulation disabilities, lack of stamina, difficulty interpreting and reacting to sensory information, and extremes of physical size. The intent of these sections is to allow a person with a physical disability to independently get to, enter, and use a site, facility, building, or element.

Section 1004 of this standard provides criteria for Type B units. These criteria are intended to be consistent with the intent of the criteria of the U.S. Department of Housing and Urban Development (HUD) Fair Housing Accessibility Guidelines. The Type B units are intended to supplement, not replace, Accessible units or Type A units as specified in this standard.

This standard is intended for adoption by government agencies and by organizations setting model codes to achieve uniformity in the technical design criteria in building codes and other regulations.

101.1 Applicability. Sites, facilities, buildings, and elements required to be accessible shall comply with the applicable provisions of Chapters 3 through 9.

EXCEPTIONS:

1. Accessible units shall comply with Section 1002.

2. Type A units shall comply with Section 1003.

3. Type B units shall comply with Section 1004.

4. Dwelling units and sleeping units required to have accessible communication features shall comply with Section 1005.

102 Anthropometric Provisions

The technical criteria in this standard are based on adult dimensions and anthropometrics. This standard also contains technical criteria based on children's dimensions and anthropometrics for drinking fountains, water closets, toilet compartments, lavatories and sinks, dining surfaces and work surfaces.

103 Compliance Alternatives

Nothing in this standard is intended to prevent the use of designs, products, or technologies as alternatives to those prescribed by this standard, provided they result in equivalent or greater accessibility and such equivalency is approved by the administrative authority adopting this standard.

104 Conventions

104.1 General. Where specific criteria of this standard differ from the general criteria of this standard, the specific criteria shall apply.

104.2 Dimensions. Dimensions that are not stated as "maximum" or "minimum" are absolute. All dimensions are subject to conventional industry tolerances.

104.3 Figures. Unless specifically stated, figures included herein are provided for informational purposes only and are not considered part of the standard.

104.4 Floor or Floor Surface. The terms floor or floor surface refer to the finish floor surface or ground surface, as applicable.

104.5 Referenced Sections. Unless specifically stated otherwise, a reference to another section or subsection within this standard includes all subsections of the referenced section or subsection.

105 Referenced Standards

105.1 General. The standards listed in Section 105.2 shall be considered part of this standard to the prescribed extent of each such reference. Where criteria in this standard differ from those of these referenced standards, the criteria of this standard shall apply.

105.2 Standards.

105.2.1 Manual on Uniform Traffic Control Devices: MUTCD - 2000 (The Federal Highway Administration, Office of Transportation Operations, Room 3408, 400 7th Street, S.W., Washington, DC 20590)

Convention	Description
36 / 915	dimension showing English units (in inches unless otherwise specified) above the line and SI units (in millimeters unless otherwise specified) below the line
6 / 150	dimension for small measurements
33 – 36 / 840 – 915	dimension showing a range with minimum – maximum
min	minimum
max	maximum
>	greater than
≥	greater than or equal to
<	less than
≤	less than or equal to
- - - - - - - - -	boundary of clear floor space or maneuvering clearance
— - —— - —— ℄	centerline
- —— - —— - ——	a permitted element or its extension
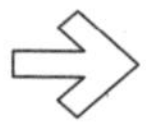	direction of travel or approach
	a wall, floor, ceiling or other element cut in section or plan
	a highlighted element in elevation or plan
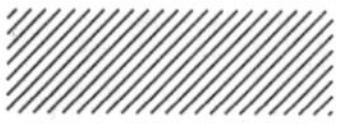	location zone of element, control or feature

Fig. 104.2
Graphic Convention for Figures

105.2.2 National Fire Alarm Code: NFPA 72-2002 (National Fire Protection Association, 1 Batterymarch Park, Quincy, MA 02269-9101)

105.2.3 Power Assist and Low Energy Power Operated Doors: ANSI/BHMA A156.19-1997 (Builders Hardware Manufacturers' Association, 355 Lexington Avenue, 17th Floor, New York, NY 10017)

105.2.4 Power Operated Pedestrian Doors: ANSI/BHMA A156.10-1999 (Builders Hardware Manufacturers' Association, 355 Lexington Avenue, 17th Floor, New York, NY 10017)

105.2.5 Safety Code for Elevators and Escalators: ASME/ANSI A17.1-2000 and Addenda A17.1a-2002 (American Society of Mechanical Engineers International, Three Park Avenue, New York, NY 10016-5990)

105.2.6 Safety Standard for Platform Lifts and Stairway Chairlifts: ASME/ANSI A18.1-1999, with Addenda A18.1a-2001 and A18.1b - 2001 (American Society of Mechanical Engineers International, Three Park Avenue, New York, NY 10016-5990)

106 Definitions

106.1 General. For the purpose of this standard, the terms listed in Section 106.5 have the indicated meaning.

106.2 Terms Defined in Referenced Standards. Terms specifically defined in a referenced standard, and not defined in this section, shall have the specified meaning from the referenced standard.

106.3 Undefined Terms. The meaning of terms not specifically defined in this standard or in a referenced standard shall be as defined by collegiate dictionaries in the sense that the context implies.

106.4 Interchangeability. Words, terms, and phrases used in the singular include the plural, and those used in the plural include the singular.

106.5 Defined Terms.

accessible: Describes a site, building, facility, or portion thereof that complies with this standard.

administrative authority: A jurisdictional body that adopts or enforces regulations and standards for the design, construction, or operation of buildings and facilities.

characters: Letters, numbers, punctuation marks, and typographic symbols.

children's use: Spaces and elements specifically designed for use primarily by people 12 years old and younger.

circulation path: An exterior or interior way of passage from one place to another for pedestrians.

counter slope: Any slope opposing the running slope of a curb ramp.

cross slope: The slope that is perpendicular to the direction of travel (see running slope).

curb ramp: A short ramp cutting through a curb or built up to it.

destination-oriented elevator system: An elevator system that provides lobby controls for the selection of destination floors, lobby indicators designating which elevator to board, and a car indicator designating the floors at which the car will stop.

detectable warning: A standardized surface feature built in or applied to floor surfaces to warn of hazards on a circulation path.

dwelling unit: A single unit providing complete, independent living facilities for one or more persons including permanent provisions for living, sleeping, eating, cooking and sanitation.

element: An architectural or mechanical component of a building, facility, space, or site.

elevator car call sequential step scanning: A technology used to enter a car call by means of an up or down floor selection button.

facility: All or any portion of a building, structure, or area, including the site on which such building, structure, or area is located, wherein specific services are provided or activities are performed.

key surface: The surface or plane of any key or button that must be touched to activate or deactivate an operable part or a machine function or enter data.

marked crossing: A crosswalk or other identified path intended for pedestrian use in crossing a vehicular way.

operable part: A component of an element used to insert or withdraw objects, or to activate, deactivate, or adjust the element.

pictogram: A pictorial symbol that represents activities, facilities, or concepts.

ramp: A walking surface that has a running slope steeper than 1:20.

running slope: The slope that is parallel to the direction of travel (see cross slope).

sign: An architectural element composed of displayed textual, symbolic, tactile, or pictorial information.

site: A parcel of land bounded by a property line or a designated portion of a public right-of-way.

sleeping unit: A room or space in which people sleep that can also include permanent provisions for living, sleeping, eating, and either sanitation or kitchen facilities but not both. Such rooms and spaces that are also part of a dwelling unit are not sleeping units.

tactile: Describes an object that can be perceived using the sense of touch.

TTY: An abbreviation for teletypewriter. Equipment that employs interactive, text-based communications through the transmission of coded signals across the standard telephone network. The term TTY also refers to devices known as text telephones and TDDs.

vehicular way: A route provided for vehicular traffic.

walk: An exterior pathway with a prepared surface for pedestrian use.

Chapter 2. Scoping

201 General

This standard provides technical criteria for making sites, facilities, buildings, and elements accessible. The administrative authority shall provide scoping provisions to specify the extent to which these technical criteria apply. These scoping provisions shall address the application of this standard to: each building and occupancy type; new construction, alterations, temporary facilities, and existing buildings; specific site and building elements; and to multiple elements or spaces provided within a site or building.

202 Dwelling and Sleeping Units

Chapter 10 of this standard contains dwelling unit and sleeping unit criteria for Accessible units, Type A units, Type B units, and units with accessible communication features. The administrative authority shall specify, in separate scoping provisions, the extent to which these technical criteria apply. These scoping provisions shall address the types and numbers of units required to comply with each set of unit criteria.

203 Administration

The administrative authority shall provide an appropriate review and approval process to ensure compliance with this standard.

Chapter 3. Building Blocks

301 General

301.1 Scope. The provisions of Chapter 3 shall apply where required by the scoping provisions adopted by the administrative authority or by Chapters 4 through 10.

302 Floor Surfaces

302.1 General. Floor surfaces shall be stable, firm, and slip resistant, and shall comply with Section 302. Changes in level in floor surfaces shall comply with Section 303.

302.2 Carpet. Carpet or carpet tile shall be securely attached and shall have a firm cushion, pad, or backing or no cushion or pad. Carpet or carpet tile shall have a level loop, textured loop, level cut pile, or level cut/uncut pile texture. The pile shall be $1/2$ inch (13 mm) maximum in height. Exposed edges of carpet shall be fastened to the floor and shall have trim along the entire length of the exposed edge. Carpet edge trim shall comply with Section 303.

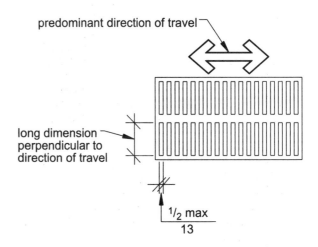

Fig. 302.3
Openings in Floor Surfaces

Fig. 303.2
Carpet on Floor Surfaces

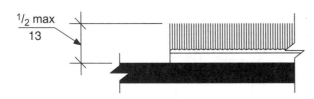

Fig. 302.2
Carpet on Floor Surfaces

302.3 Openings. Openings in floor surfaces shall be of a size that does not permit the passage of a $1/2$ inch (13 mm) diameter sphere, except as allowed in Sections 407.4.3, 408.4.3, 409.4.3, 410.4, and 805.10. Elongated openings shall be placed so that the long dimension is perpendicular to the dominant direction of travel.

303 Changes in Level

303.1 General. Changes in level in floor surfaces shall comply with Section 303.

303.2 Vertical. Changes in level of $1/4$ inch (6.4 mm) maximum in height shall be permitted to be vertical.

303.3 Beveled. Changes in level greater than $1/4$ inch (6.4 mm) in height and not more than $1/2$ inch (13 mm) maximum in height shall be beveled with a slope not steeper than 1:2.

Changes in level greater than $1/2$ inch (13 mm) in height shall be ramped and shall comply with Section 405 or 406.

304 Turning Space

304.1 General. A turning space shall comply with Section 304.

304.2 Floor Surface. Floor surfaces of a turning space shall have a slope not steeper than 1:48 and shall comply with Section 302.

304.3 Size. Turning spaces shall comply with Section 304.3.1 or 304.3.2.

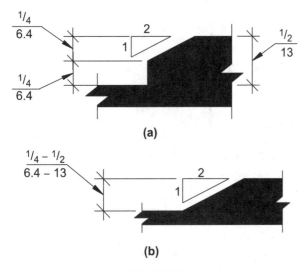

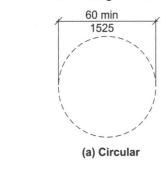

Fig. 303.3
Beveled Changes in Level

(a) Circular

(b) T-shaped

Fig. 304.3
Size of Turning Space

304.3.1 Circular Space. The turning space shall be a circular space with a 60-inch (1525 mm) minimum diameter. The turning space shall be permitted to include knee and toe clearance complying with Section 306.

304.3.2 T-Shaped Space. The turning space shall be a T-shaped space within a 60-inch (1525 mm) minimum square, with arms and base 36 inches (915 mm) minimum in width. Each arm of the T shall be clear of obstructions 12 inches (305 mm) minimum in each direction, and the base shall be clear of obstructions 24 inches (610 mm) minimum. The turning space shall be permitted to include knee and toe clearance complying with Section 306 only at the end of either the base or one arm.

304.4 Door Swing. Unless otherwise specified, doors shall be permitted to swing into turning spaces.

305 Clear Floor Space

305.1 General. A clear floor space shall comply with Section 305.

305.2 Floor Surfaces. Floor surfaces of a clear floor space shall have a slope not steeper than 1:48 and shall comply with Section 302.

305.3 Size. The clear floor space shall be 48 inches (1220 mm) minimum in length and 30 inches (760 mm) minimum in width.

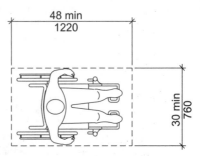

Fig. 305.3
Size of Clear Floor Space

305.4 Knee and Toe Clearance. Unless otherwise specified, clear floor space shall be permitted to include knee and toe clearance complying with Section 306.

305.5 Position. Unless otherwise specified, the clear floor space shall be positioned for either forward or parallel approach to an element.

305.6 Approach. One full, unobstructed side of the clear floor space shall adjoin or overlap an accessible route or adjoin another clear floor space.

305.7 Alcoves. If a clear floor space is in an alcove or otherwise confined on all or part of three sides, additional maneuvering clearances complying with Sections 305.7.1 and 305.7.2 shall be provided, as applicable.

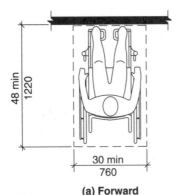

(a) Forward

(b) Parallel

Fig. 305.5
Position of Clear Floor Space

305.7.1 Parallel Approach. Where the clear floor space is positioned for a parallel approach, the alcove shall be 60 inches (1525 mm) minimum in width where the depth exceeds 15 inches (380 mm).

305.7.2 Forward Approach. Where the clear floor space is positioned for a forward approach, the alcove shall be 36 inches (915 mm) minimum in width where the depth exceeds 24 inches (610 mm).

306 Knee and Toe Clearance

306.1 General. Where space beneath an element is included as part of clear floor space at an element, clearance at an element, or a turning space, the space shall comply with Section 306. Additional space beyond knee and toe clearance shall be permitted beneath elements.

306.2 Toe Clearance.

306.2.1 General. Space beneath an element between the floor and 9 inches (230 mm) above the floor shall be considered toe clearance and shall comply with Section 306.2.

306.2.2 Maximum Depth. Toe clearance shall be permitted to extend 25 inches (635 mm) maximum under an element.

306.2.3 Minimum Depth. Where toe clearance is required at an element as part of a clear floor space, the toe clearance shall extend 17 inches (430 mm) minimum beneath the element.

306.2.4 Additional Clearance. Space extending greater than 6 inches (150 mm) beyond the available knee clearance at 9 inches (230 mm) above the floor shall not be considered toe clearance.

306.2.5 Width. Toe clearance shall be 30 inches (760 mm) minimum in width.

306.3 Knee Clearance.

306.3.1 General. Space beneath an element between 9 inches (230 mm) and 27 inches (685 mm) above the floor shall be considered knee clearance and shall comply with Section 306.3.

306.3.2 Maximum Depth. Knee clearance shall be permitted to extend 25 inches (635 mm) maximum under an element at 9 inches (230 mm) above the floor.

306.3.3 Minimum Depth. Where knee clearance is required beneath an element as part of a clear floor space, the knee clearance shall be 11

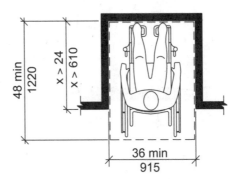

(a) Forward Approach

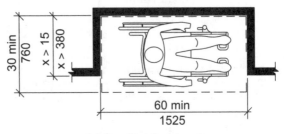

(b) Parallel Approach

Fig. 305.7
Maneuvering Clearance in an Alcove

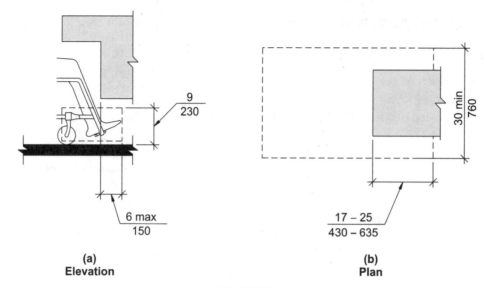

(a)
Elevation

(b)
Plan

Fig. 306.2
Toe Clearance

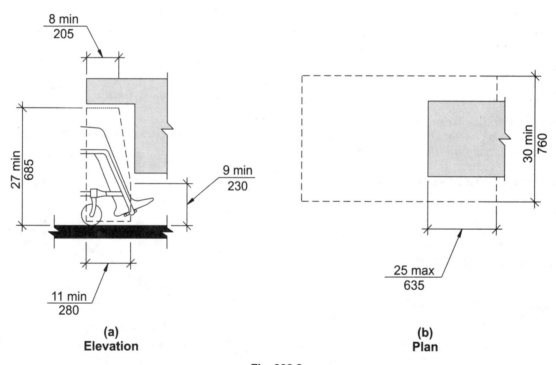

(a)
Elevation

(b)
Plan

Fig. 306.3
Knee Clearance

inches (280 mm) minimum in depth at 9 inches (230 mm) above the floor, and 8 inches (205 mm) minimum in depth at 27 inches (685 mm) above the floor.

306.3.4 Clearance Reduction. Between 9 inches (230 mm) and 27 inches (685 mm) above the floor, the knee clearance shall be permitted to be reduced at a rate of 1 inch (25 mm) for each 6 inches (150 mm) in height.

306.3.5 Width. Knee clearance shall be 30 inches (760 mm) minimum in width.

307 Protruding Objects

307.1 General. Protruding objects on circulation paths shall comply with Section 307.

307.2 Protrusion Limits. Objects with leading edges more than 27 inches (685 mm) and not more than 80 inches (2030 mm) above the floor shall protrude 4 inches (100 mm) maximum horizontally into the circulation path.

EXCEPTIONS:

1. Handrails shall be permitted to protrude 4 $1/2$ inches (115 mm) maximum.

2. Door closers and door stops shall be permitted to be 78 inches (1980 mm) minimum above the floor.

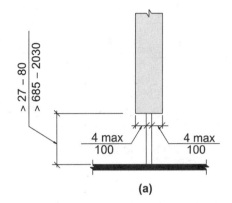

(a)

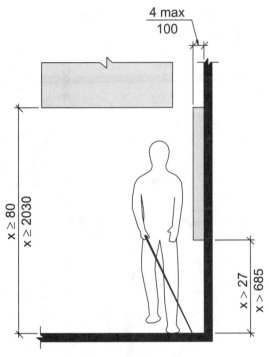

Fig. 307.2
Limits of Protruding Objects

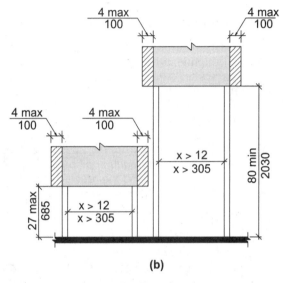

(b)

Fig. 307.3
Post-Mounted Protruding Objects

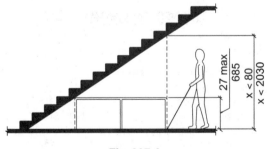

Fig. 307.4
Reduced Vertical Clearance

307.3 Post-Mounted Objects. Objects on posts or pylons shall be permitted to overhang 4 inches (100 mm) maximum where more than 27 inches (685 mm) and not more than 80 inches (2030 mm) above the floor. Objects on multiple posts or pylons where the clear distance between the posts or pylons is greater than 12 inches (305 mm) shall have the lowest edge of such object either 27 inches (685 mm) maximum or 80 inches (2030 mm) minimum above the floor.

307.4 Reduced Vertical Clearance. Guardrails or other barriers shall be provided where object protrusion is beyond the limits allowed by Sections 307.2 and 307.3, and where the vertical clearance is less than 80 inches (2030 mm) above the floor. The leading edge of such guardrail or barrier shall be 27 inches (685 mm) maximum above the floor.

307.5 Required Clear Width. Protruding objects shall not reduce the clear width required for accessible routes.

308 Reach Ranges

308.1 General. Reach ranges shall comply with Section 308.

308.2 Forward Reach.

308.2.1 Unobstructed. Where a forward reach is unobstructed, the high forward reach shall be 48 inches (1220 mm) maximum and the low forward reach shall be 15 inches (380 mm) minimum above the floor.

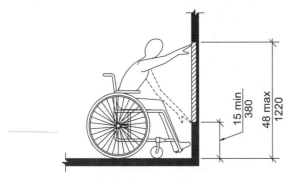

Fig. 308.2.1
Unobstructed Forward Reach

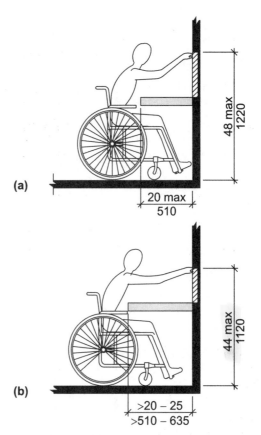

(a)

(b)

Fig. 308.2.2
Obstructed High Forward Reach

308.2.2 Obstructed High Reach. Where a high forward reach is over an obstruction, the clear floor space shall extend beneath the element for a distance not less than the required reach depth over the obstruction. The high forward reach shall be 48 inches (1220 mm) maximum where the reach depth is 20 inches (510 mm) maximum. Where the reach depth exceeds 20 inches (510 mm), the high forward reach shall be 44 inches (1120 mm) maximum, and the reach depth shall be 25 inches (635 mm) maximum.

308.3 Side Reach.

308.3.1 Unobstructed. Where a clear floor space allows a parallel approach to an element and the side reach is unobstructed, the high side reach shall be 48 inches (1220 mm) maximum and the low side reach shall be 15 inches (380 mm) minimum above the floor.

> **EXCEPTION:** Existing elements shall be permitted at 54 inches (1370 mm) maximum above the floor.

308.3.2 Obstructed High Reach. Where a clear floor space allows a parallel approach to an object and the high side reach is over an obstruction, the height of the obstruction shall be 34 inches (865 mm) maximum and the depth of the obstruction shall be 24 inches (610 mm) maximum. The high side reach shall be 48 inches (1220 mm) maximum for a reach depth of 10 inches (255 mm) maximum. Where the reach depth exceeds 10 inches (255 mm), the high side reach shall be 46 inches (1170 mm) maximum for a reach depth of 24 inches (610 mm) maximum.

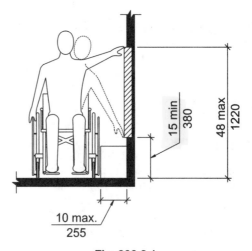

Fig. 308.3.1
Unobstructed Side Reach

309 Operable Parts

309.1 General. Operable parts required to be accessible shall comply with Section 309.

309.2 Clear Floor Space. A clear floor space complying with Section 305 shall be provided.

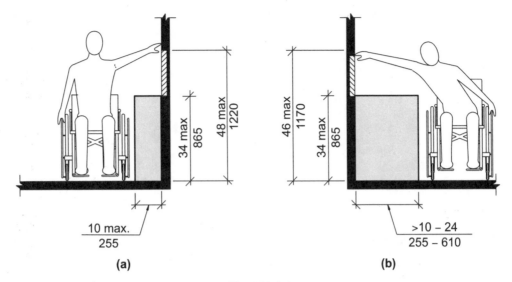

Fig. 308.3.2
Obstructed High Side Reach

309.3 Height. Operable parts shall be placed within one or more of the reach ranges specified in Section 308.

309.4 Operation. Operable parts shall be operable with one hand and shall not require tight grasping, pinching, or twisting of the wrist. The force required to activate operable parts shall be 5.0 pounds (22.2 N) maximum.

EXCEPTION: Gas pump nozzles shall not be required to provide operable parts that have an activating force of 5.0 pounds (22.2 N) maximum.

Chapter 4. Accessible Routes

401 General

401.1 Scope. Accessible routes required by the scoping provisions adopted by the administrative authority shall comply with the applicable provisions of Chapter 4.

402 Accessible Routes

402.1 General. Accessible routes shall comply with Section 402.

402.2 Components. Accessible routes shall consist of one or more of the following components: Walking surfaces with a slope not steeper than 1:20, doors and doorways, ramps, curb ramps excluding the flared sides, elevators, and platform lifts. All components of an accessible route shall comply with the applicable portions of this standard.

402.3 Revolving Doors, Revolving Gates, and Turnstiles. Revolving doors, revolving gates, and turnstiles shall not be part of an accessible route.

403 Walking Surfaces

403.1 General. Walking surfaces that are a part of an accessible route shall comply with Section 403.

403.2 Floor Surface. Floor surfaces shall comply with Section 302.

403.3 Slope. The running slope of walking surfaces shall not be steeper than 1:20. The cross slope of a walking surface shall not be steeper than 1:48.

403.4 Changes in Level. Changes in level shall comply with Section 303.

403.5 Clear Width. Clear width of an accessible route shall comply with Table 403.5.

Table 403.5—Clear Width of an Accessible Route

Segment Length	Minimum Segment Width
≤ 24 inches (610 mm)	32 inches (815 mm)[1]
> 24 inches (610 mm)	36 inches (915 mm)

[1]Consecutive segments of 32 inches (815 mm) in width must be separated by a route segment 48 inches (1220 mm) minimum in length and 36 inches (915 mm) minimum in width.

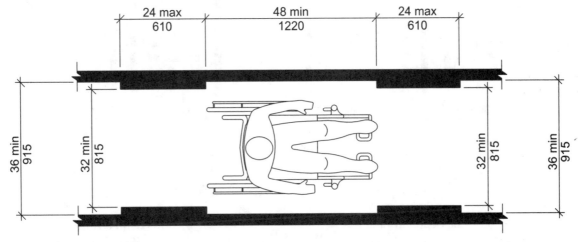

Fig. 403.5
Clear Width of an Accessible Route

403.5.1 Clear Width at Turn. Where an accessible route makes a 180 degree turn around an object that is less than 48 inches (1220 mm) in width, clear widths shall be 42 inches (1065 mm) minimum approaching the turn, 48 inches (1220 mm) minimum during the turn, and 42 inches (1065 mm) minimum leaving the turn.

> **EXCEPTION:** Section 403.5.1 shall not apply where the clear width at the turn is 60 inches (1525 mm) minimum.

403.5.2 Passing Space. An accessible route with a clear width less than 60 inches (1525 mm) shall provide passing spaces at intervals of 200 feet (61 m) maximum. Passing spaces shall be either a 60 inch (1525 mm) minimum by 60 inch (1525 mm) minimum space, or an intersection of two walking surfaces that provide a T-shaped turning space complying with Section 304.3.2, provided the base and arms of the T-shaped space extend 48 inches (1220 mm) minimum beyond the intersection.

403.6 Handrails. Where handrails are required at the side of a corridor they shall comply with Sections 505.4 through 505.9.

404 Doors and Doorways

404.1 General. Doors and doorways that are part of an accessible route shall comply with Section 404.

404.2 Manual Doors. Manual doors and doorways, and manual gates, including ticket gates, shall comply with the requirements of Section 404.2.

> **EXCEPTION:** Doors, doorways, and gates designed to be operated only by security personnel shall not be required to comply with Sections 404.2.6, 404.2.7, and 404.2.8.

404.2.1 Double-Leaf Doors and Gates. At least one of the active leaves of doorways with two leaves shall comply with Sections 404.2.2 and 404.2.3.

404.2.2 Clear Width. Doorways shall have a clear opening width of 32 inches (815 mm) minimum. Clear opening width of doorways with swinging doors shall be measured between the face of door and stop, with the door open 90 degrees. Openings, doors and doorways without doors more than 24 inches (610 mm) in depth shall provide a clear opening width of 36 inches (915 mm) minimum. There shall be no projections into the clear opening width lower than 34

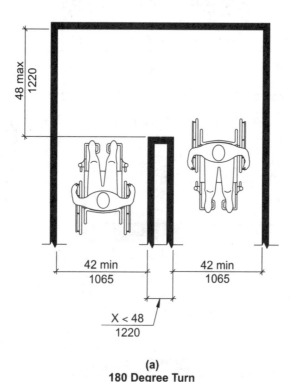

(a)
180 Degree Turn

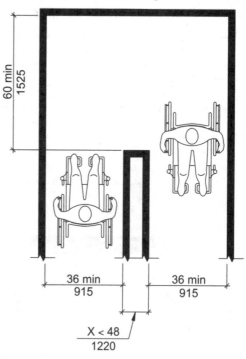

(b)
**180 Degree Turn
(Exception)**

**Fig. 403.5.1
Clear Width at Turn**

inches (865 mm) above the floor. Projections into the clear opening width between 34 inches (865 mm) and 80 inches (2030 mm) above the floor shall not exceed 4 inches (100 mm).

EXCEPTIONS:

1. Door closers and door stops shall be permitted to be 78 inches (1980 mm) minimum above the floor.

2. In alterations, a projection of $^5/_8$ inch (16 mm) maximum into the required clear opening width shall be permitted for the latch side stop.

404.2.3 Maneuvering Clearances at Doors. Minimum maneuvering clearances at doors shall comply with Section 404.2.3 and shall include the full clear opening width of the doorway.

404.2.3.1 Swinging Doors. Swinging doors shall have maneuvering clearances complying with Table 404.2.3.1.

404.2.3.2 Sliding and Folding Doors. Sliding doors and folding doors shall have maneuvering clearances complying with Table 404.2.3.2.

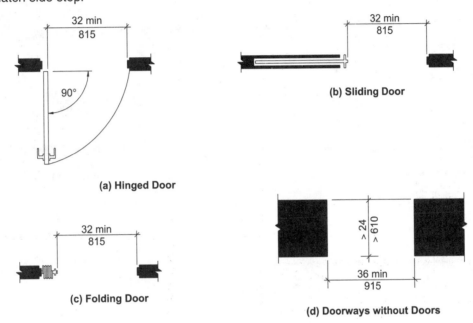

(a) Hinged Door

(b) Sliding Door

(c) Folding Door

(d) Doorways without Doors

Fig. 404.2.2
Clear Width of Doorways

Table 404.2.3.1—Maneuvering Clearances at Manual Swinging Doors

TYPE OF USE		MINIMUM MANEUVERING CLEARANCES	
Approach Direction	Door Side	Perpendicular to Doorway	Parallel to Doorway (beyond latch unless noted)
From front	Pull	60 inches (1525 mm)	18 inches (455 mm)
From front	Push	48 inches (1220 mm)	0 inches (0 mm)[3]
From hinge side	Pull	60 inches (1525 mm)	36 inches (915 mm)
From hinge side	Pull	54 inches (1370 mm)	42 inches (1065 mm)
From hinge side	Push	42 inches (1065 mm)[1]	22 inches (560 mm)[3 & 4]
From latch side	Pull	48 inches (1220 mm)[2]	24 inches (610 mm)
From latch side	Push	42 inches (1065 mm)[2]	24 inches (610 mm)

[1]Add 6 inches (150 mm) if closer and latch provided.
[2]Add 6 inches (150 mm) if closer provided.
[3]Add 12 inches (305 mm) beyond latch if closer and latch provided.
[4]Beyond hinge side.

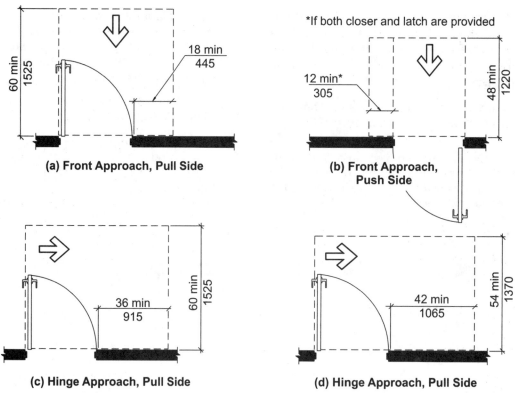

(a) Front Approach, Pull Side

*If both closer and latch are provided

(b) Front Approach, Push Side

(c) Hinge Approach, Pull Side

(d) Hinge Approach, Pull Side

* If both closer and latch are provided
** 48 min (1220) if both closer and latch provided

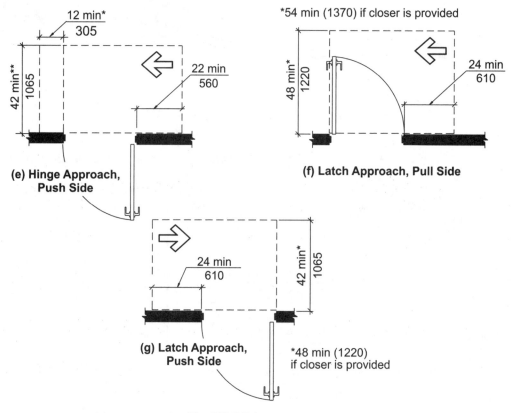

(e) Hinge Approach, Push Side

*54 min (1370) if closer is provided

(f) Latch Approach, Pull Side

(g) Latch Approach, Push Side

*48 min (1220) if closer is provided

Fig. 404.2.3.1
Maneuvering Clearance at Manual Swinging Doors

404.2.3.3 Doorways without Doors. Doorways without doors that are less than 36 inches (915 mm) in width shall have maneuvering clearances complying with Table 404.2.3.3

404.2.3.4 Recessed Doors. Where any obstruction within 18 inches (455 mm) of the latch side of a doorway projects more than 8 inches (205 mm) beyond the face of the door, measured perpendicular to the face of the door, maneuvering clearances for a forward approach shall be provided.

404.2.3.5 Floor Surface. Floor surface within the maneuvering clearances shall have a slope not steeper than 1:48 and shall comply with Section 302.

404.2.4 Thresholds at Doorways. If provided, thresholds at doorways shall be $1/2$ inch (13 mm) maximum in height. Raised thresholds and changes in level at doorways shall comply with Sections 302 and 303.

EXCEPTION: Section 404.2.4 shall not apply to existing thresholds or altered thresholds $3/4$ inch (19 mm) maximum in height that have a beveled edge on each side with a maximum slope of 1:2 for the height exceeding $1/4$ inch (6.4 mm).

404.2.5 Two Doors in Series. Distance between two hinged or pivoted doors in series shall be 48 inches (1220 mm) minimum plus the width of any door swinging into the space. The space between the doors shall provide a turning space complying with Section 304.

Table 404.2.3.2—Maneuvering Clearances at Sliding and Folding Doors

Approach Direction	MINIMUM MANEUVERING CLEARANCES	
	Perpendicular to Doorway	**Parallel to Doorway (beyond stop or latch side unless noted)**
From front	48 inches (1220 mm)	0 inches (0 mm)
From nonlatch side	42 inches (1065 mm)	22 inches (560 mm)[1]
From latch side	42 inches (1065 mm)	24 inches (610 mm)

[1]Beyond pocket or hinge side.

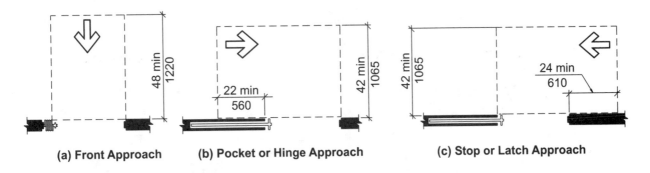

(a) Front Approach   **(b) Pocket or Hinge Approach**   **(c) Stop or Latch Approach**

Fig. 404.2.3.2
Maneuvering Clearance at Sliding and Folding Doors

Table 404.2.3.3—Maneuvering Clearances for Doorways without Doors

Approach Direction	MINIMUM MANEUVERING CLEARANCES Perpendicular to Doorway
From front	48 inches (1220 mm)
From side	42 inches (1065 mm)

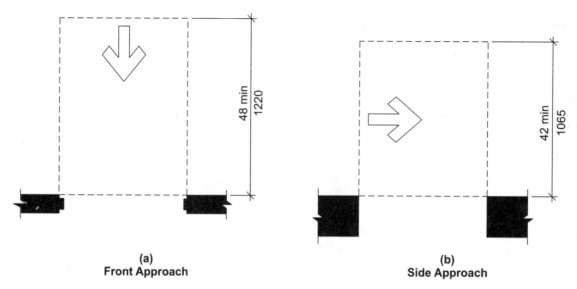

(a)
Front Approach

(b)
Side Approach

Fig. 404.2.3.3
Maneuvering Clearance at Doorways without Doors

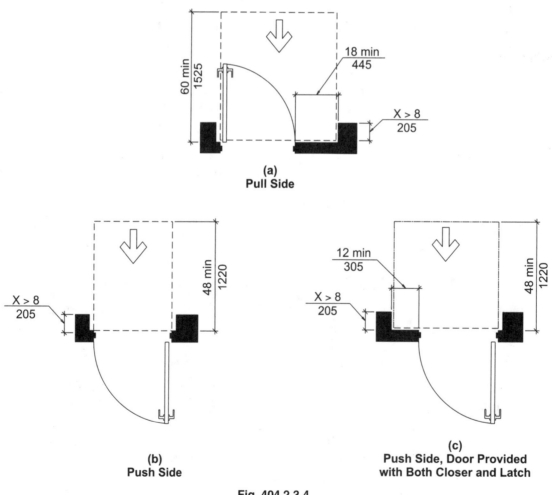

(a)
Pull Side

(b)
Push Side

(c)
Push Side, Door Provided
with Both Closer and Latch

Fig. 404.2.3.4
Maneuvering Clearance at Recessed Doors

404.2.6 Door Hardware. Handles, pulls, latches, locks, and other operable parts on accessible doors shall have a shape that is easy to grasp with one hand and does not require tight grasping, pinching, or twisting of the wrist to operate. Operable parts of such hardware shall be 34 inches (865 mm) minimum and 48 inches (1220 mm) maximum above the floor. Where sliding doors are in the fully open position, operating hardware shall be exposed and usable from both sides.

EXCEPTION: Locks used only for security purposes and not used for normal operation are permitted in any location.

404.2.7 Closing Speed.

404.2.7.1 Door Closers. Door closers shall be adjusted so that from an open position of 90 degrees, the time required to move the door to an open position of 12 degrees shall be 5 seconds minimum.

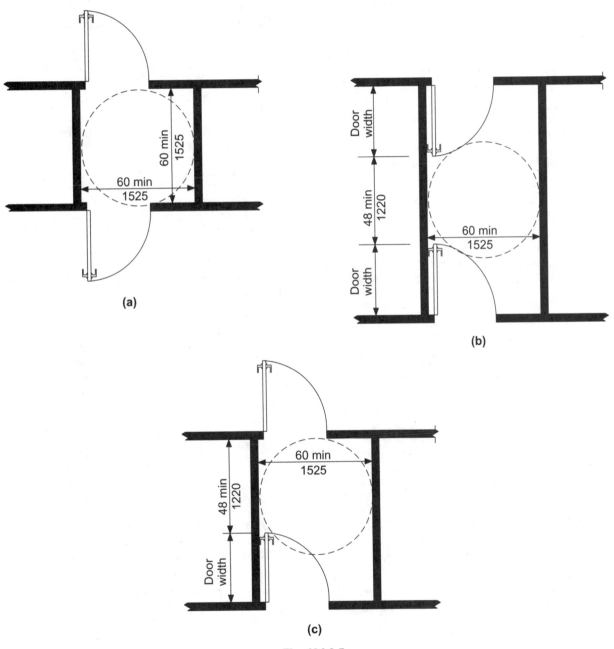

(a)

(b)

(c)

Fig. 404.2.5
Two Doors in a Series

404.2.7.2 Spring Hinges. Door spring hinges shall be adjusted so that from the open position of 70 degrees, the door shall move to the closed position in 1.5 seconds minimum.

404.2.8 Door-Opening Force. Fire doors shall have the minimum opening force allowable by the appropriate administrative authority. The force for pushing or pulling open doors other than fire doors shall be as follows:

1. Interior hinged door: 5.0 pounds (22.2 N) maximum

2. Sliding or folding door: 5.0 pounds (22.2 N) maximum

These forces do not apply to the force required to retract latch bolts or disengage other devices that hold the door in a closed position.

404.2.9 Door Surface. Door surfaces within 10 inches (255 mm) of the floor, measured vertically, shall be a smooth surface on the push side extending the full width of the door. Parts creating horizontal or vertical joints in such surface shall be within $\frac{1}{16}$ inch (1.6 mm) of the same plane as the other. Cavities created by added kick plates shall be capped.

EXCEPTIONS:

1. Sliding doors.

2. Tempered glass doors without stiles and having a bottom rail or shoe with the top leading edge tapered at no less than 60 degrees from the horizontal shall not be required to meet the 10 inch (255 mm) bottom rail height requirement.

3. Doors that do not extend to within 10 inches (255 mm) of the floor.

404.2.10 Vision Lites. Doors and sidelites adjacent to doors containing one or more glazing panels that permit viewing through the panels shall have the bottom of at least one panel on either the door or an adjacent sidelite 43 inches (1090 mm) maximum above the floor.

EXCEPTION: Vision lites with the lowest part more than 66 inches (1675 mm) above the floor are not required to comply with Section 404.2.10.

404.3 Automatic Doors. Automatic doors and automatic gates shall comply with Section 404.3. Full powered automatic doors shall comply with ANSI/BHMA A156.10 listed in Section 105.2.4.

Power-assist and low-energy doors shall comply with ANSI/BHMA A156.19 listed in Section 105.2.3.

EXCEPTION: Doors, doorways, and gates designed to be operated only by security personnel shall not be required to comply with Sections 404.3.2, 404.3.4, and 404.3.5.

404.3.1 Clear Opening Width. Doorways shall have a clear opening width of 32 inches (815 mm) in power-on and power-off mode. The minimum clear opening width for automatic door systems shall be based on the clear opening width provided with all leafs in the open position.

404.3.2 Maneuvering Clearances. Maneuvering clearances at power-assisted doors shall comply with Section 404.2.3.

404.3.3 Thresholds. Thresholds and changes in level at doorways shall comply with Section 404.2.4.

404.3.4 Two Doors in Series. Doors in series shall comply with Section 404.2.5.

404.3.5 Control Switches. Manually operated control switches shall comply with Section 309. The clear floor space adjacent to the control switch shall be located beyond the arc of the door swing.

405 Ramps

405.1 General. Ramps along accessible routes shall comply with Section 405.

405.2 Slope. Ramp runs shall have a running slope not steeper than 1:12.

EXCEPTION: In existing buildings or facilities, ramps shall be permitted to have slopes steeper than 1:12 complying with Table 405.2 where such slopes are necessary due to space limitations.

405.3 Cross Slope. Cross slope of ramp runs shall not be steeper than 1:48.

405.4 Floor Surfaces. Floor surfaces of ramp runs shall comply with Section 302.

405.5 Clear Width. The clear width of a ramp run shall be 36 inches (915 mm) minimum. Where handrails are provided on the ramp run, the clear width shall be measured between the handrails.

405.6 Rise. The rise for any ramp run shall be 30 inches (760 mm) maximum.

Table 405.2—Allowable Ramp Dimensions for Construction in Existing Sites, Buildings, and Facilities

Slope[1]	Maximum Rise
Steeper than 1:10 but not steeper than 1:8	3 inches (75 mm)
Steeper than 1:12 but not steeper than 1:10	6 inches (150 mm)

[1]A slope steeper than 1:8 shall not be permitted.

405.7 Landings. Ramps shall have landings at bottom and top of each ramp run. Landings shall comply with Section 405.7.

405.7.1 Slope. Landings shall have a slope not steeper than 1:48 and shall comply with Section 302.

405.7.2 Width. Clear width of landings shall be at least as wide as the widest ramp run leading to the landing.

405.7.3 Length. Landings shall have a clear length of 60 inches (1525 mm) minimum.

405.7.4 Change in Direction. Ramps that change direction at ramp landings shall be sized to provide a turning space complying with Section 304.3.

405.7.5 Doorways. Where doorways are adjacent to a ramp landing, maneuvering clearances required by Sections 404.2.3 and 404.3.2 shall be permitted to overlap the landing area. Where doors that are subject to locking are adjacent to a ramp landing, landings shall be sized to provide a turning space complying with Section 304.3.

405.8 Handrails. Ramp runs with a rise greater than 6 inches (150 mm) shall have handrails complying with Section 505.

405.9 Edge Protection. Edge protection complying with Section 405.9.1 or 405.9.2 shall be provided on each side of ramp runs and at each side of ramp landings.

EXCEPTIONS:

1. Ramps not required to have handrails where curb ramp flares complying with Section 406.3 are provided.

2. Sides of ramp landings serving an adjoining ramp run or stairway.

3. Sides of ramp landings having a vertical drop-off of $1/2$ inch (13 mm) maximum within 10 inches (255 mm) horizontally of the minimum landing area.

405.9.1 Extended Floor Surface. The floor surface of the ramp run or ramp landing shall extend 12 inches (305 mm) minimum beyond the inside face of a railing complying with Section 505.

405.9.2 Curb or Barrier. A curb or barrier shall be provided that prevents the passage of a 4-inch (100 mm) diameter sphere where any portion of the sphere is within 4 inches (100 mm) of the floor.

405.10 Wet Conditions. Landings subject to wet conditions shall be designed to prevent the accumulation of water.

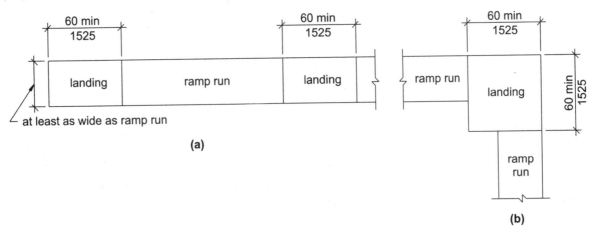

Fig. 405.7
Ramp Landings

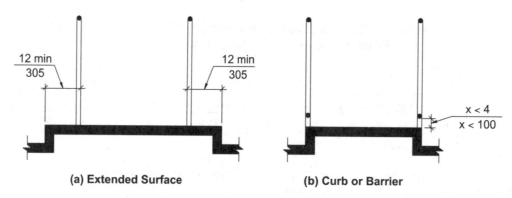

(a) Extended Surface

(b) Curb or Barrier

Fig. 405.9
Ramp Edge Protection

406 Curb Ramps

406.1 General. Curb ramps on accessible routes shall comply with Sections 406, 405.2, 405.3, and 405.10.

406.2 Counter Slope. Counter slopes of adjoining gutters and road surfaces immediately adjacent to the curb ramp shall not be steeper than 1:20. The adjacent surfaces at transitions at curb ramps to walks, gutters and streets shall be at the same level.

406.3 Sides of Curb Ramps. Where provided, curb ramp flares shall not be steeper than 1:10.

406.4 Width. Curb ramps shall be 36 inches (915 mm) minimum in width, exclusive of flared sides.

406.5 Floor Surface. Floor surfaces of curb ramps shall comply with Section 302.

406.6 Location. Curb ramps and the flared sides of curb ramps shall be located so they do not project into vehicular traffic lanes, parking spaces, or parking access aisles. Curb ramps at marked crossings shall be wholly contained within the markings, excluding any flared sides.

406.7 Landings. Landings shall be provided at the tops of curb ramps. The clear length of the landing shall be 36 inches (915 mm) minimum. The clear

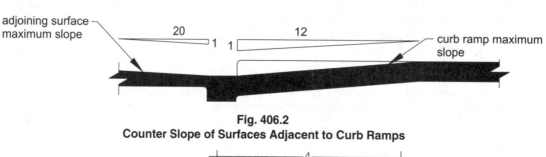

Fig. 406.2
Counter Slope of Surfaces Adjacent to Curb Ramps

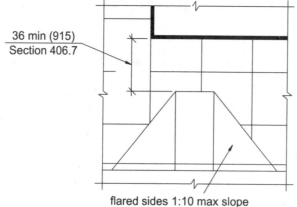

Fig. 406.3
Sides of Curb Ramps

width of the landing shall be at least as wide as the curb ramp, excluding flared sides, leading to the landing.

>   **EXCEPTION:** In alterations, where there is no landing at the top of curb ramps, curb ramp flares shall be provided and shall not be steeper than 1:12.

406.8 Obstructions. Curb ramps shall be located or protected to prevent their obstruction by parked vehicles.

406.9 Handrails. Handrails are not required on curb ramps.

406.10 Diagonal Curb Ramps. Diagonal or corner-type curb ramps with returned curbs or other well-defined edges shall have the edges parallel to the direction of pedestrian flow. The bottoms of diagonal curb ramps shall have 48 inches (1220 mm) minimum clear space outside active traffic lanes of the roadway. Diagonal curb ramps provided at marked crossings shall provide the 48 inches (1220 mm) minimum clear space within the markings. Diagonal curb ramps with flared sides shall have a segment of curb 24 inches (610 mm) minimum in length on each side of the curb ramp and within the marked crossing.

406.11 Islands. Raised islands in crossings shall be a cut-through level with the street or have curb ramps at both sides. Each curb ramp shall have a level area 48 inches (1220 mm) minimum in length and 36 inches (915 mm) minimum in width at the top of the curb ramp in the part of the island intersected by the crossings. Each 48-inch (1220 mm) by 36-inch (915 mm) area shall be oriented so the 48-inch (1220 mm) length is in the direction of the running slope of the curb ramp it serves. The 48-inch (1220 mm) by 36-inch (915 mm) areas and the accessible route shall be permitted to overlap.

406.12 Detectable Warnings at Raised Marked Crossings. Marked crossings that are raised to the same level as the adjoining sidewalk shall be preceded by a 24-inch (610 mm) deep detectable warning complying with Section 705, extending the full width of the marked crossing.

406.13 Detectable Warnings at Curb Ramps. Where detectable warnings are provided on curb ramps, they shall comply with Sections 406.13 and 705.

>   **406.13.1 Area Covered.** Detectable warnings shall be 24 inches (610 mm) minimum in the direction of travel and extend the full width of the curb ramp or flush surface.

>   **406.13.2 Location.** The detectable warning shall be located so the edge nearest the curb line is 6 inches (150 mm) to 8 inches (205 mm) from the curb line.

406.14 Detectable Warnings at Islands or Cut-through Medians. Where detectable warnings are provided on curb ramps or at raised marked crossings leading to islands or cut-through medians, the island or cut-through median shall also be provided with detectable warnings complying with Section 705, are 24 inches (610 mm) in depth, and extend the full width of the pedestrian route or cut-through. Where such island or cut-through median is less than 48 inches (1220 mm) in depth,

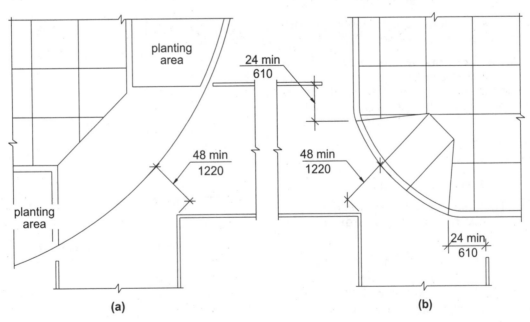

**Fig. 406.10
Diagonal Curb Ramps**

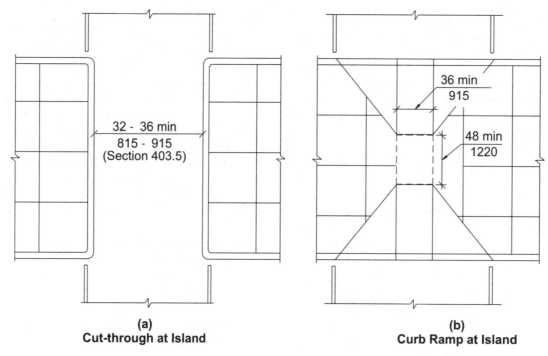

(a)
Cut-through at Island

(b)
Curb Ramp at Island

Fig. 406.11
Islands

the entire width and depth of the pedestrian route or cut-through shall have detectable warnings.

407 Elevators

407.1 General. Elevators shall comply with Section 407 and ASME A17.1 listed in Section 105.2.5. Elevators shall be passenger elevators as classified by ASME A17.1. Elevator operation shall be automatic.

407.2 Elevator Landing Requirements. Elevator landings shall comply with Section 407.2.

407.2.1 Call Controls. Where elevator call buttons or keypads are provided, they shall comply with Sections 407.2.1 and 309.4. Call buttons shall be raised or flush. Objects beneath hall call buttons shall protrude 1 inch (25 mm) maximum.

EXCEPTIONS:

1. Existing elevators shall be permitted to have recessed call buttons.

2. The restriction on objects beneath call buttons shall not apply to existing call buttons.

407.2.1.1 Height. Call buttons and keypads shall be located within one of the reach ranges specified in Section 308, measured to the centerline of the highest operable part.

EXCEPTION: Existing call buttons and existing keypads shall be permitted to be

located 54 inches (1370 mm) maximum above the floor, measured to the centerline of the highest operable part.

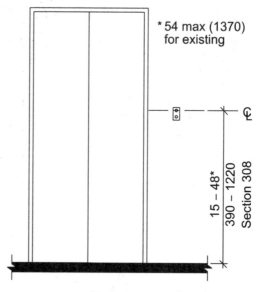

Fig. 407.2.1.1
Height of Elevator Call Buttons

407.2.1.2 Size. Call buttons shall be $^3/_4$ inch (19 mm) minimum in the smallest dimension.

EXCEPTION: Existing elevator call buttons shall not be required to comply with Section 407.2.1.2.

407.2.1.3 Clear Floor Space. A clear floor space complying with Section 305 shall be provided at call controls.

407.2.1.4 Location. The call button that designates the up direction shall be located above the call button that designates the down direction.

> **EXCEPTION:** Destination-oriented elevators shall not be required to comply with Section 407.2.1.4.

407.2.1.5 Signals. Call buttons shall have visible signals to indicate when each call is registered and when each call is answered.

> **EXCEPTIONS:**
>
> 1. Destination-oriented elevators shall not be required to comply with Section 407.2.1.5, provided visible and audible signals complying with Section 407.2.1.7 are provided.
>
> 2. Existing elevators shall not be required to comply with Section 407.2.1.5.

407.2.1.6 Keypads. Where keypads are provided, keypads shall be in a standard telephone keypad arrangement and shall comply with Section 407.4.7.2.

407.2.1.7 Destination-oriented Elevator Signals. Destination-oriented elevators shall be provided with visible and audible signals to indicate which car is responding to a call. The audible signal shall be activated by pressing a function button. The function button shall be identified by the International Symbol for Accessibility and tactile indication. The International Symbol for Accessibility, complying with Section 703.6.3.1, shall be $5/_8$ inch (16 mm) in height and be a visual character complying with Section 703.2. The tactile indication shall be three raised dots, spaced $1/_4$ inch (6.4 mm) at base diameter, in the form of an equilateral triangle. The function button shall be located immediately below the keypad arrangement or floor buttons.

407.2.2 Hall Signals. Hall signals, including in-car signals, shall comply with Section 407.2.2.

407.2.2.1 Visible and Audible Signals. A visible and audible signal shall be provided at each hoistway entrance to indicate which car is answering a call and the car's direction of travel. Where in-car signals are provided they shall be visible from the floor area adjacent to the hall call buttons.

> **EXCEPTIONS:**
>
> 1. Destination-oriented elevators shall not be required to comply with Section 407.2.2.1, provided visible and audible signals complying with Section 407.2.1.7 are provided.
>
> 2. In existing elevators, a signal indicating the direction of car travel shall not be required.

407.2.2.2 Visible Signals. Visible signal fixtures shall be centered at 72 inches (1830 mm) minimum above the floor. The visible signal elements shall be $2^1/_2$ inches (64 mm) minimum measured along the vertical centerline

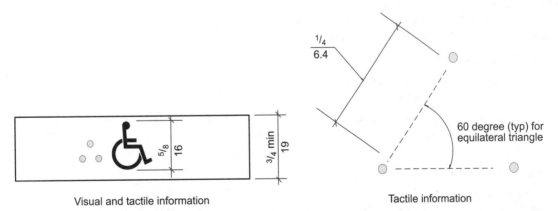

Visual and tactile information

Tactile information

Fig. 407.2.1.7
Destination-oriented Elevator Indication

of the element. Signals shall be visible from the floor area adjacent to the hall call button.

EXCEPTIONS:

1. Destination-oriented elevators shall be permitted to have signals visible from the floor area adjacent to the hoistway entrance.

2. Existing elevators shall not be required to comply with Section 407.2.2.2.

407.2.2.3 Audible Signals. Audible signals shall sound once for the up direction and twice for the down direction, or shall have verbal annunciators that indicate the direction of elevator car travel. Audible signals shall have a frequency of 1500 Hz maximum. Verbal annunciators shall have a frequency of 300 Hz minimum and 3,000 Hz maximum. The audible signal or verbal annunciator shall be 10 dBA minimum above ambient, but shall not exceed 80 dBA, measured at the hall call button.

EXCEPTIONS:

1. Destination-oriented elevators shall not be required to comply with Section 407.2.2.3, provided the audible tone and verbal announcement is the same as those given at the call button or call button keypad.

2. The requirement for the frequency and range of audible signals shall not apply in existing elevators.

407.2.2.4 Differentiation. Each destination-oriented elevator in a bank of elevators shall have audible and visible means for differentiation.

407.2.3 Hoistway Signs. Signs at elevator hoistways shall comply with Section 407.2.3.

407.2.3.1 Floor Designation. Floor designations shall be provided in tactile characters complying with Section 703.3 located on both jambs of elevator hoistway entrances. Tactile characters shall be 2 inches (51 mm) minimum in height. A tactile star shall be provided on both jambs at the main entry level.

407.2.3.2 Car Designations. Destination-oriented elevators shall provide car identification in tactile characters complying with Section 703.3 located on both jambs of the hoistway immediately below the floor designation. Tactile characters shall be 2 inches (51 mm) minimum in height.

407.2.4 Destination Signs. Where signs indicate that elevators do not serve all landings, signs in tactile characters complying with Section 703.3 shall be provided above the hall call button fixture.

EXCEPTION: Destination oriented elevator systems shall not be required to comply with Section 407.2.4.

407.3 Elevator Door Requirements. Hoistway and elevator car doors shall comply with Section 407.3.

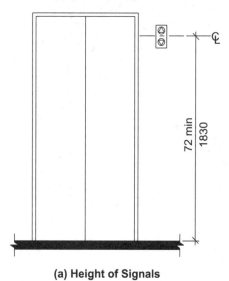

(a) Height of Signals

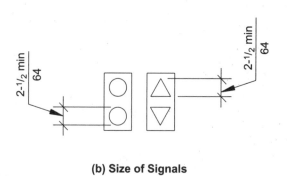

(b) Size of Signals

Fig. 407.2.2.2
Elevator Visible Signals

407.3.1 Type. Elevator doors shall be horizontal sliding type. Car gates shall be prohibited.

407.3.2 Operation. Elevator hoistway and car doors shall open and close automatically.

> **EXCEPTION:** Existing manually operated hoistway swing doors shall be permitted, provided:
>
> a) they comply with Sections 404.2.2 and 404.2.8;
>
> b) the car door closing is not initiated until the hoistway door is closed.

407.3.3 Reopening Device. Elevator doors shall be provided with a reopening device complying with Section 407.3.3 that shall stop and reopen a car door and hoistway door automatically if the door becomes obstructed by an object or person.

> **EXCEPTION:** In existing elevators, manually operated doors shall not be required to comply with Section 407.3.3.

407.3.3.1 Height. The reopening device shall be activated by sensing an obstruction passing through the opening at 5 inches (125 mm) nominal and 29 inches (735 mm) nominal above the floor.

407.3.3.2 Contact. The reopening device shall not require physical contact to be activated, although contact shall be permitted before the door reverses.

407.3.3.3 Duration. The reopening device shall remain effective for 20 seconds minimum.

407.3.4 Door and Signal Timing. The minimum acceptable time from notification that a car is answering a call until the doors of that car start to close shall be calculated from the following equation:

$T = D/(1.5 \text{ ft/s})$ or $T = D/(455 \text{ mm/s}) = 5$ seconds minimum, where T equals the total time in seconds and D equals the distance (in feet or millimeters) from the point in the lobby or corridor 60 inches (1525 mm) directly in front of the farthest call button controlling that car to the centerline of its hoistway door.

> **EXCEPTIONS:**
>
> 1. For cars with in-car lanterns, T shall be permitted to begin when the signal is visible from the point 60 inches (1525 mm) directly in front of the farthest hall call button and the audible signal is sounded.
>
> 2. Destination-oriented elevators shall not be required to comply with Section 407.3.4.

407.3.5 Door Delay. Elevator doors shall remain fully open in response to a car call for 3 seconds minimum.

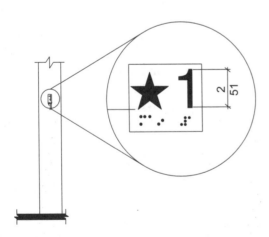

Fig. 407.2.3.1
Floor Designation

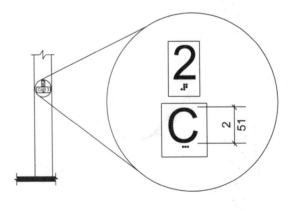

Fig. 407.2.3.2
Designation-oriented Elevator Car Identification

407.3.6 Width. Elevator door clear opening width shall comply with Table 407.4.1.

EXCEPTION: In existing elevators, a power-operated car door complying with Section 404.2.2 shall be permitted.

407.4 Elevator Car Requirements. Elevator cars shall comply with Section 407.4.

407.4.1 Car Dimensions. Inside dimensions of elevator cars shall comply with Table 407.4.1.

EXCEPTION: Existing elevator car configurations that provide a clear floor area of 16 square feet (1.5 m²) minimum, and provide a clear inside dimension of 36 inches (915 mm) minimum in width and 54 inches (1370 mm) minimum in depth, shall be permitted.

407.4.2 Floor Surfaces. Floor surfaces in elevator cars shall comply with Section 302.

407.4.3 Platform to Hoistway Clearance. The clearance between the car platform sill and the edge of any hoistway landing shall be in compliance with ASME/ANSI A17.1 listed in Section 105.2.5.

407.4.4 Leveling. Each car shall be equipped with a self-leveling feature that will automatically bring and maintain the car at floor landings within a tolerance of $1/_2$ inch (13 mm) under rated loading to zero loading conditions.

407.4.5 Illumination. The level of illumination at the car controls, platform, car threshold and car landing sill shall be 5 foot-candles (54 lux) minimum.

407.4.6 Elevator Car Controls. Where provided, elevator car controls shall comply with Sections 407.4.6 and 309.

EXCEPTION: In existing elevators, where a new car operating panel complying with Section 407.4.6 is provided, existing car operating panels shall not be required to comply with Section 407.4.6.

407.4.6.1 Location. Controls shall be located within one of the reach ranges specified in Section 308.

EXCEPTIONS:

1. Where the elevator panel serves more than 16 openings and a parallel approach to the controls is provided, buttons with floor designations shall be permitted to be 54 inches (1370 mm) maximum above the floor.

2. In existing elevators, where a parallel approach is provided to the controls, car control buttons with floor designations shall be permitted to be located 54 inches (1370 mm) maximum above the floor. Where the panel is changed, it shall comply with Section 407.4.6.1.

Table 407.4.1—Minimum Dimensions of Elevator Cars[1]

Door Location	Door Clear Opening Width	Inside Car, Side to Side	Inside Car, Back Wall to Front Return	Inside Car, Back Wall to Inside Face of Door
Centered	42 inches (1065 mm)	80 inches (2030 mm)	51 inches (1295 mm)	54 inches (1370 mm)
Side (Off Center)	36 inches (915 mm)[1]	68 inches (1725 mm)	51 inches (1295 mm)	54 inches (1370 mm)
Any	36 inches (915 mm)[1]	54 inches (1370 mm)	80 inches (2030 mm)	80 inches (2030 mm)
Any	36 inches (915 mm)[1]	60 inches (1525 mm)[2]	60 inches (1525 mm)[2]	60 inches (1525 mm)[2]

[1] A tolerance of minus $5/_8$ inch (16 mm) is permitted.

[2] Other car configurations that provide a 36-inch (915 mm) door clear opening width and a turning space complying with Section 304 with the door closed are permitted.

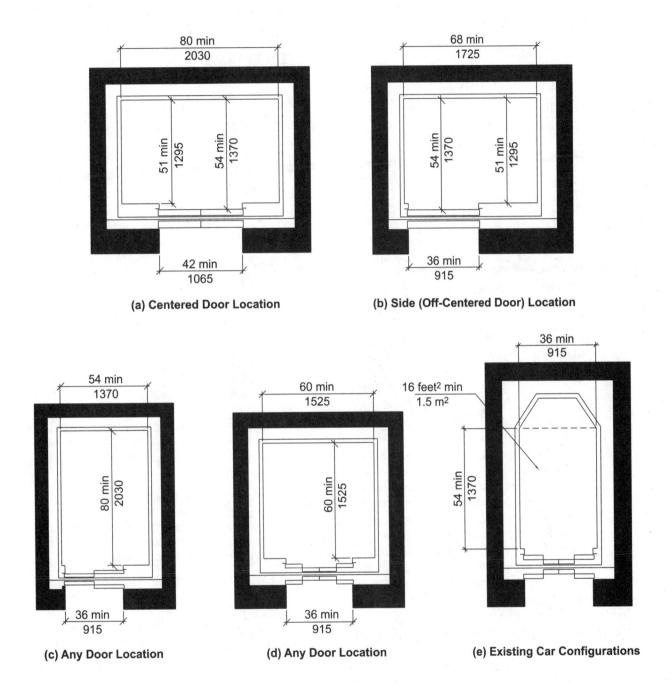

Fig. 407.4.1
Inside Dimensions of Elevator Cars

407.4.6.2 Buttons. Car control buttons with floor designations shall be raised or flush, and shall comply with Section 407.4.6.2.

EXCEPTION: In existing elevators, buttons shall be permitted to be recessed.

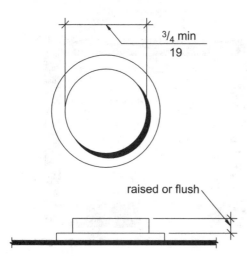

Fig. 407.4.6.2
Elevator Car Control Buttons

407.4.6.2.1 Size. Buttons shall be $^3/_4$ inch (19 mm) minimum in their smallest dimension.

407.4.6.2.2 Arrangement. Buttons shall be arranged with numbers in ascending order. Floors shall be designated . . . -4, -3, -2, -1, 0, 1, 2, 3, 4, et cetera, with floors below the main entry floor designated with minus numbers. Numbers shall be permitted to be omitted, provided the remaining numbers are in sequence. Where a telephone keypad arrangement is used, the number key ("#") shall be utilized to enter the minus symbol ("-"). When two or more columns of buttons are provided they shall read from left to right.

407.4.6.3 Keypads. Car control keypads shall be in a standard telephone keypad arrangement and shall comply with Section 407.4.7.2.

407.4.6.4 Emergency Controls. Emergency controls shall comply with Section 407.4.6.4.

407.4.6.4.1 Height. Emergency control buttons shall have their centerlines 35 inches (890 mm) minimum above the floor.

407.4.6.4.2 Location. Emergency controls, including the emergency alarm, shall be grouped at the bottom of the panel.

407.4.7 Designations and Indicators of Car Controls. Designations and indicators of car controls shall comply with Section 407.4.7.

EXCEPTIONS:

1. In existing elevators, where a new car operating panel complying with Section 407.4.7 is provided, existing car operating panels shall not be required to comply with Section 407.4.7.

2. Where existing building floor designations differ from the arrangement required by Section 407.4.6.2.2, or are alphanumeric, a new operating panel shall be permitted to use such existing building floor designations.

407.4.7.1 Buttons. Car control buttons shall comply with Section 407.4.7.1.

407.4.7.1.1 Type. Control buttons shall be identified by tactile characters complying with Section 703.3.

407.4.7.1.2 Location. Tactile character and braille designations shall be placed immediately to the left of the control button to which the designations apply. Where a negative number is used to indicate a negative floor, the braille designation shall be a cell with the dots 3 and 6 followed by the ordinal number.

EXCEPTION: Where space on an existing car operating panel precludes tactile markings to the left of the control button, markings shall be placed as near to the control button as possible.

407.4.7.1.3 Symbols. The control button for the emergency stop, alarm, door open, door close, main entry floor, and phone, shall be identified with tactile symbols as shown in Table 407.4.7.1.3.

407.4.7.1.4 Visible Indicators. Buttons with floor designations shall be provided with visible indicators to show that a call has been registered. The visible indication shall extinguish when the car arrives at the designated floor.

407.4.7.2 Keypads. Keypads shall be identified by visual characters complying with Section 703.2 and shall be centered on the corresponding keypad button. The number five key shall have a single raised dot. The dot shall have a base diameter of 0.118 inch (3 mm) minimum to 0.120 inch (3.05 mm) maximum, and a height of 0.025 inch (0.6 mm) minimum to 0.037 inch (0.9 mm) maximum.

407.4.8 Elevator Car Call Sequential Step Scanning. Elevator car call sequential step scanning shall be provided where car control buttons are provided more than 48 inches (1220 mm) above the floor, as permitted by Section 407.4.6.1, Exception #1. Floor selection shall be accomplished by applying momentary or constant pressure to the up or down scan button. The up scan button shall sequentially select floors above the current floor. The down scan button shall sequentially select floors below the current floor. When pressure is removed from the up or down scan button for more than 2 seconds, the last floor selected shall be registered as a car call. The up and down scan button shall be located adjacent to or immediately above the emergency control buttons.

407.4.9 Car Position Indicators. Audible and visible car position indicators shall be provided in elevator cars.

407.4.9.1 Visible Indicators. Visible indicators shall comply with Section 407.4.9.1.

407.4.9.1.1 Size. Characters shall be $^1/_2$ inch (13 mm) minimum in height.

407.4.9.1.2 Location. Indicators shall be located above the car control panel or above the door.

407.4.9.1.3 Floor Arrival. As the car passes a floor and when a car stops at a floor served by the elevator, the corresponding character shall illuminate.

> **EXCEPTION:** Destination-oriented elevators shall not be required to comply with Section 407.4.9.1.3, provided the visible indicators extinguish when the call has been answered.

407.4.9.1.4 Destination Indicator. In destination-oriented elevators, a display shall be provided in the car with visible indicators to show car destinations.

407.4.9.2 Audible Indicators. Audible indicators shall comply with Section 407.4.9.2.

407.4.9.2.1 Signal Type. The signal shall be an automatic verbal annunciator that announces the floor at which the car is about to stop. The verbal announcement indicating the floor shall be completed prior to the initiation of the door opening.

> **EXCEPTION:** For elevators other than destination-oriented elevators that have a rated speed of 200 feet per minute (1 m/s) or less, a non-verbal audible signal with a frequency of 1500 Hz maximum that sounds as the car passes or is about to stop at a floor served by the elevator shall be permitted.

407.4.9.2.2 Signal Level. The verbal annunciator shall be 10 dBA minimum above ambient, but shall not exceed 80 dBA, measured at the annunciator.

407.4.9.2.3 Frequency. The verbal annunciator shall have a frequency of 300 Hz minimum to 3,000 Hz maximum.

407.4.10 Emergency Communications. Emergency two-way communication systems between the elevator car and a point outside the hoistway shall comply with Section 407.4.10 and ASME/ANSI A17.1 listed in Section 105.2.5.

407.4.10.1 Height. The highest operable part of a two-way communication system shall comply with Section 308.

407.4.10.2 Identification. Tactile characters complying with Section 703.3 and symbols complying with Section 407.4.7.1.3 shall be provided adjacent to the device.

408 Limited-Use/Limited-Application Elevators

408.1 General. Limited-use/limited-application elevators shall comply with Section 408 and ASME A17.1 listed in Section 105.2.5. Elevator operation shall be automatic.

408.2 Elevator Landing Requirements. Landings serving limited-use/limited application elevators shall comply with Section 408.2.

408.2.1 Call Controls. Elevator call buttons and keypads shall comply with Section 407.2.1.

408.2.2 Hall Signals. Hall signals shall comply with Section 407.2.2.

408.2.3 Hoistway Signs. Signs at elevator hoistways shall comply with Section 407.2.3.

408.3 Elevator Door Requirements. Elevator hoistway doors shall comply with Section 408.3.

Table 407.4.7.1.3—Control Button Identification

Control Button	Tactile Symbol	Braille Message	Proportions Open circles indicate unused dots within each Braille Cell
DOOR OPEN 3.0 mm TYP. BETWEEN ELEMENTS 16.0 mm 4.8 mm		OP"EN"	2.0 mm 2.0 mm
REAR/SIDE DOOR OPEN		REAR/SIDE OP"EN"	
DOOR CLOSE		CLOSE	
REAR/SIDE DOOR CLOSE		REAR/SIDE CLOSE	
MAIN		MA"IN"	
ALARM		AL"AR"M	
PHONE		PH"ONE"	
EMERGENCY STOP (WHEN PROVIDED) X on face of octagon is not required to be tactile		"ST"OP	

34

408.3.1 Sliding Doors. Sliding hoistway and car doors shall comply with Sections 407.3.1 through 407.3.3, and 408.3.3.

408.3.2 Swinging Doors. Swinging hoistway doors shall open and close automatically and shall comply with Sections 408.3.2, 404, and 407.3.2.

408.3.2.1 Power Operation. Swinging doors shall be power-operated and shall comply with ANSI/BHMA A156.19 listed in Section 105.2.3.

408.3.2.2 Duration. Power-operated swinging doors shall remain open for 20 seconds minimum when activated.

408.3.3 Door Location and Width. Car doors shall provide a clear opening width of 32 inches (815 mm) minimum. Car doors shall be positioned at a narrow end of the car.

EXCEPTION: Car doors that provide a clear opening width of 36 inches (915 mm) minimum shall be permitted to be located on adjacent sides of cars that provide a clear floor area of 51 inches (1295 mm) in width and 51 inches (1295 mm) in depth.

408.4 Elevator Car Requirements. Elevator cars shall comply with Section 408.4.

408.4.1 Inside Dimensions of Elevator Cars. Elevator cars shall provide a clear floor area of 42 inches (1065 mm) minimum in width, and 54 inches (1370 mm) minimum in depth.

EXCEPTIONS:

1. Cars that provide a 51 inches (1295 mm) minimum clear floor width shall be permitted to provide 51 inches (1295 mm) minimum clear floor depth.

2. For installations in existing buildings, elevator cars that provide a clear floor area of 15 square feet (1.4 m²) minimum, and provide a clear inside dimension of 36 inches (915 mm) minimum in width and 54 inches (1370 mm) minimum in depth, shall be permitted.

408.4.2 Floor Surfaces. Floor surfaces in elevator cars shall comply with Section 302.

408.4.3 Platform to Hoistway Clearance. The clearance between the car platform sill and the edge of any hoistway landing shall be in compliance with ASME/ANSI A17.1 listed in Section 105.2.5.

408.4.4 Leveling. Elevator car leveling shall comply with Section 407.4.4.

408.4.5 Illumination. Elevator car illumination shall comply with Section 407.4.5.

408.4.6 Elevator Car Controls. Elevator car controls shall comply with Section 407.4.6. Control panels shall be centered on a side wall.

408.4.7 Designations and Indicators of Car Controls. Designations and indicators of car controls shall comply with Section 407.4.7.

408.4.8 Emergency Communications. Car emergency signaling devices complying with Section 407.4.10 shall be provided.

409 Private Residence Elevators

409.1 General. Private residence elevators shall comply with Section 409 and ASME/ANSI A17.1 listed in Section 105.2.5. Elevator operation shall be automatic.

EXCEPTION: Elevators complying with Section 407 or 408.

409.2 Call Buttons. Call buttons at elevator landings shall comply with Section 309. Call buttons shall be ³/₄ inch (19 mm) minimum in their smallest dimension.

409.3 Doors and Gates. Elevator car and hoistway doors and gates shall comply with Sections 409.3 and 404.

EXCEPTION: The maneuvering clearances required by Section 404.2.3 shall not apply for approaches to the push side of swinging doors.

409.3.1 Power Operation. Elevator car doors and gates shall be power operated and shall comply with ANSI/BHMA A156.19 listed in Section 105.2.3. Elevator cars with a single opening shall have low energy power operated hoistway doors and gates.

EXCEPTION: For elevators with a car that has more than one opening, the hoistway doors and gates shall be permitted to be of the manual-open, self-close type.

409.3.2 Duration. Power operated doors and gates shall remain open for 20 seconds minimum when activated.

409.3.3 Door or Gate Location. Car gates or doors shall be positioned at a narrow end of the clear floor area required by Section 409.4.1.

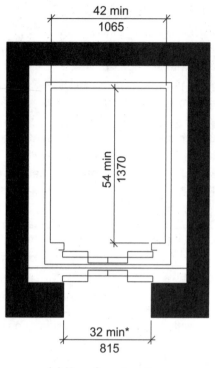

(a) New Construction

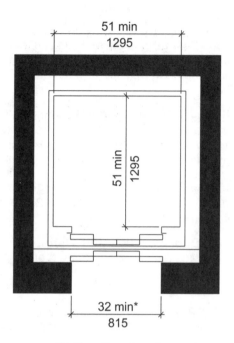

**(b) New Construction
Exception 1**

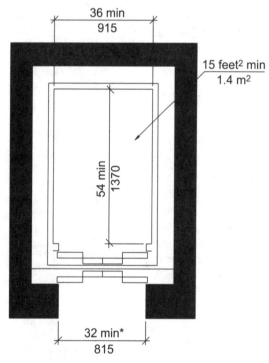

**(c) Existing Building
Exception 2**

*Door opening size from Section 408.3.3

**Fig. 408.4.1
Inside Dimensions of Limited Use/Limited Application (LULA) Elevator Cars**

409.4 Elevator Car Requirements. Elevator cars shall comply with Section 409.4.

409.4.1 Inside Dimensions of Elevator Cars. Elevator cars shall provide a clear floor area 36 inches (915 mm) minimum in width and 48 inches (1220 mm) minimum in depth.

409.4.2 Floor Surfaces. Floor surfaces in elevator cars shall comply with Section 302.

409.4.3 Platform to Hoistway Clearance. The clearance between the car platform sill and the edge of any hoistway landing shall be $1^1/_4$ inches (32 mm) maximum.

409.4.4 Leveling. Each car shall automatically stop at a floor landing within a tolerance of $^1/_2$ inch (13 mm) under rated loading to zero loading conditions.

409.4.5 Illumination. The level of illumination at the car controls, platform, and car threshold and landing sill shall be 5 foot-candles (54 lux) minimum.

409.4.6 Elevator Car Controls. Elevator car controls shall comply with Sections 409.4.6 and 309.4.

409.4.6.1 Buttons. Control buttons shall be $^3/_4$ inch (19 mm) minimum in their smallest dimension. Control buttons shall be raised or flush.

409.4.6.2 Height. Buttons with floor designations shall comply with Section 309.3.

409.4.6.3 Location. Controls shall be on a sidewall, 12 inches (305 mm) minimum from any adjacent wall.

409.4.7 Emergency Communications. Emergency communications systems shall comply with Section 409.4.7.

409.4.7.1 Type. A telephone and emergency signal device shall be provided in the car.

409.4.7.2 Operable Parts. The telephone and emergency signaling device shall comply with Section 309.3.

409.4.7.3 Compartment. If the device is in a closed compartment, the compartment door hardware shall comply with Section 309.

409.4.7.4 Cord. The telephone cord shall be 29 inches (735 mm) minimum in length.

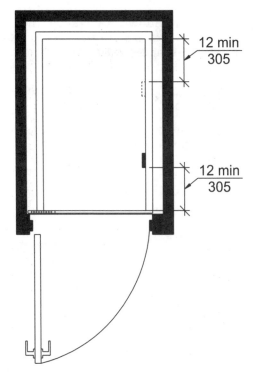

Fig. 409.4.6.3
Location of Controls in Private
Residence Elevators

410 Platform Lifts

410.1 General. Platform lifts shall comply with Section 410 and ASME/ANSI A18.1 listed in Section 105.2.6. Platform lifts shall not be attendant operated and shall provide unassisted entry and exit from the lift.

410.2 Lift Entry. Lifts with doors or gates shall comply with Section 410.2.1. Lifts with ramps shall comply with Section 410.2.2.

410.2.1 Doors and Gates. Doors and gates shall be low energy power operated doors or gates complying with Section 404.3. Doors shall remain open for 20 seconds minimum. End door clear opening width shall be 32 inches (815 mm) minimum. Side door clear opening width shall be 42 inches (1065 mm) minimum.

EXCEPTION: Lifts serving two landings maximum and having doors or gates on opposite sides shall be permitted to have self-closing manual doors or gates.

410.2.2 Ramps. End ramps shall be 32 inches (815 mm) minimum in width. Side ramps shall be 42 inches (1065 mm) minimum in width.

410.3 Floor Surfaces. Floor surfaces of platform lifts shall comply with Section 302.

410.4 Platform to Runway Clearance. The clearance between the platform sill and the edge of any runway landing shall be $1\frac{1}{4}$ inch (32 mm) maximum.

410.5 Clear Floor Space. Clear floor space of platform lifts shall comply with Section 305.

410.6 Operable Parts. Controls for platform lifts shall comply with Section 309.

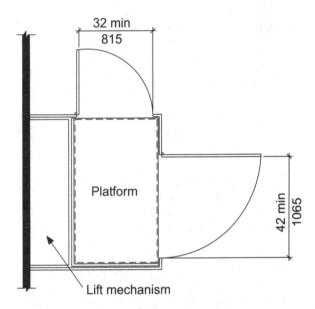

Fig. 410.2.1
Platform Lift Doors and Gates

Chapter 5. General Site and Building Elements

501 General

501.1 Scope. General site and building elements required to be accessible by the scoping provisions adopted by the administrative authority shall comply with the applicable provisions of Chapter 5.

502 Parking Spaces

502.1 General. Accessible car and van parking spaces shall comply with Section 502.

502.2 Vehicle Space Size. Car parking spaces shall be 96 inches (2440 mm) minimum in width. Van parking spaces shall be 132 inches (3350 mm) minimum in width.

> **EXCEPTION:** Van parking spaces shall be permitted to be 96 inches (2440 mm) minimum in width where the adjacent access aisle is 96 inches (2440 mm) minimum in width.

502.3 Vehicle Space Marking. Car and van parking spaces shall be marked to define the width. Where parking spaces are marked with lines, the width measurements of parking spaces and adjacent access aisles shall be made from the centerline of the markings.

> **EXCEPTION:** Where parking spaces or access aisles are not adjacent to another parking space or access aisle, measurements shall be permit-

ted to include the full width of the line defining the parking space or access aisle.

502.4 Access Aisle. Car and van parking spaces shall have an adjacent access aisle complying with Section 502.4.

502.4.1 Location. Access aisles shall adjoin an accessible route. Two parking spaces shall be permitted to share a common access aisle. Access aisles shall not overlap with the vehicular way. Parking spaces shall be permitted to have access aisles placed on either side of the car or van parking space. Van parking spaces that are angled shall have access aisles located on the passenger side of the parking space.

502.4.2 Width. Access aisles serving car and van parking spaces shall be 60 inches (1525 mm) minimum in width.

502.4.3 Length. Access aisles shall extend the full length of the parking spaces they serve.

502.4.4 Marking. Access aisles shall be marked so as to discourage parking in them. Where access aisles are marked with lines, the width measurements of access aisles and adjacent parking spaces shall be made from the centerline of the markings.

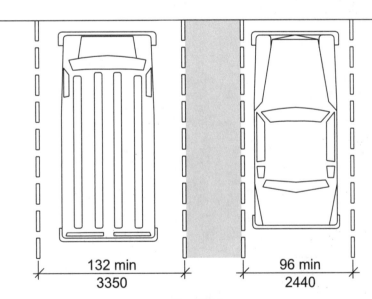

132 min
3350

96 min
2440

Fig. 502.2
Vehicle Parking Space Size

EXCEPTION: Where access aisles or parking spaces are not adjacent to another access aisle or parking space, measurements shall be permitted to include the full width of the line defining the access aisle or parking space.

502.5 Floor Surfaces. Parking spaces and access aisles shall comply with Section 302 and have surface slopes not steeper than 1:48. Access aisles shall be at the same level as the parking spaces they serve.

502.6 Vertical Clearance. Parking spaces for vans, access aisles serving them, and vehicular routes from an entrance to the van parking spaces, and from the van parking spaces to a vehicular exit serving them shall provide a vertical clearance of 98 inches (2490 mm) minimum.

502.7 Identification. Where accessible parking spaces are required to be identified by signs, the signs shall include the International Symbol of Accessibility complying with Section 703.6.3.1. Signs identifying van parking spaces shall contain the designation "van accessible." Such signs shall be 60 inches (1525 mm) minimum above the floor of the parking space, measured to the bottom of the sign.

502.8 Relationship to Accessible Routes. Parking spaces and access aisles shall be designed so that cars and vans, when parked, cannot obstruct the required clear width of adjacent accessible routes.

503 Passenger Loading Zones

503.1 General. Accessible passenger loading zones shall comply with Section 503.

503.2 Vehicle Pull-up Space Size. Passenger loading zones shall provide a vehicular pull-up space 96 inches (2440 mm) minimum in width and 20 feet (6100 mm) minimum in length.

503.3 Access Aisle. Passenger loading zones shall have an adjacent access aisle complying with Section 503.3.

503.3.1 Location. Access aisles shall adjoin an accessible route. Access aisles shall not overlap the vehicular way.

503.3.2 Width. Access aisles serving vehicle pull-up spaces shall be 60 inches (1525 mm) minimum in width.

503.3.3 Length. Access aisles shall be 20 feet (6100 mm) minimum in length.

503.3.4 Marking. Access aisles shall be marked so as to discourage parking in them.

503.4 Floor Surfaces. Vehicle pull-up spaces and access aisles serving them shall comply with Section 302 and shall have slopes not steeper than 1:48. Access aisles shall be at the same level as the vehicle pull-up space they serve.

503.5 Vertical Clearance. Vehicle pull-up spaces, access aisles serving them, and a vehicular route from an entrance to the passenger loading zone, and from the passenger loading zone to a vehicular exit serving them, shall provide a vertical clearance of 114 inches (2895 mm) minimum.

504 Stairways

504.1 General. Accessible stairs shall comply with Section 504.

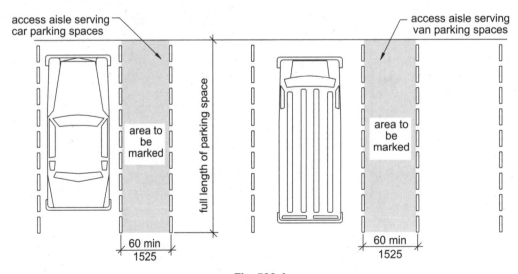

access aisle serving car parking spaces

access aisle serving van parking spaces

area to be marked

full length of parking space

area to be marked

60 min
1525

60 min
1525

Fig. 502.4
Parking Space Access Aisle

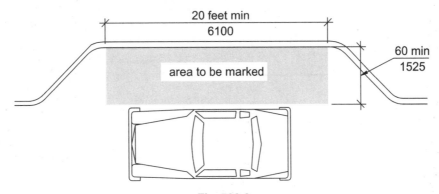

Fig. 503.3
Passenger Loading Zone Access Aisle

504.2 Treads and Risers. All steps on a flight of stairs shall have uniform riser height and uniform tread depth. Risers shall be 4 inches (100 mm) minimum and 7 inches (180 mm) maximum in height. Treads shall be 11 inches (280 mm) minimum in depth.

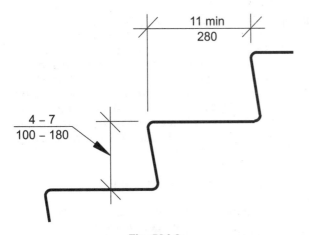

Fig. 504.2
Treads and Risers for Accessible Stairways

504.3 Open Risers. Open risers shall not be permitted.

504.4 Tread Surface. Stair treads shall comply with Section 302 and shall have a slope not steeper than 1:48.

504.5 Nosings. The radius of curvature at the leading edge of the tread shall be ¹/₂ inch (13 mm) maximum. Nosings that project beyond risers shall have the underside of the leading edge curved or beveled. Risers shall be permitted to slope under the tread at an angle of 30 degrees maximum from vertical. The permitted projection of the nosing shall be 1¹/₂ inches (38 mm) maximum over the tread or floor below. The leading 2 inches (51 mm) of the tread shall have visual contrast of dark-on-light or light-on-dark from the remainder of the tread.

504.6 Handrails. Stairs shall have handrails complying with Section 505.

504.7 Wet Conditions. Stair treads and landings subject to wet conditions shall be designed to prevent the accumulation of water.

504.8 Lighting. Lighting for interior stairways shall comply with Section 504.8.

504.8.1 Luminance Level. Lighting facilities shall be capable of providing 10 foot-candles (108 lux) of luminance measured at the center of tread surfaces and on landing surfaces within 24 inches (610 mm) of step nosings.

504.8.2 Lighting Controls. If provided, occupancy-sensing automatic controls shall activate the stairway lighting so the luminance level required by Section 504.8.1 is provided on the entrance landing, each stair flight adjacent to the entrance landing, and on the landings above and below the entrance landing prior to any step being used.

504.9 Stair Level Identification. Stair level identification signs in tactile characters complying with Section 703.3 shall be located at each floor level landing in all enclosed stairways adjacent to the door leading from the stairwell into the corridor to identify the floor level. The exit door discharging to the outside or to the level of exit discharge shall have a tactile sign stating "EXIT."

505 Handrails

505.1 General. Handrails required by Section 405.8 for ramps, or Section 504.6 for stairs, shall comply with Section 505.

505.2 Location. Handrails shall be provided on both sides of stairs and ramps.

EXCEPTION: Aisle stairs and aisle ramps provided with a handrail either at the side or within the aisle width.

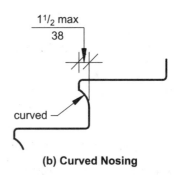

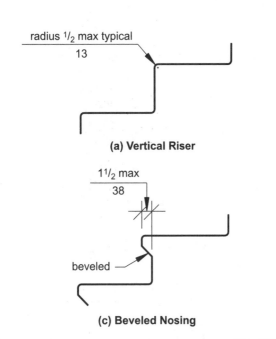

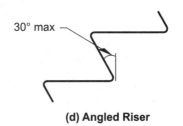

(a) Vertical Riser

(b) Curved Nosing

(c) Beveled Nosing

(d) Angled Riser

Fig. 504.5
Stair Nosings

505.3 Continuity. Handrails shall be continuous within the full length of each stair flight or ramp run. Inside handrails on switchback or dogleg stairs or ramps shall be continuous between flights or runs. Other handrails shall comply with Sections 505.10 and 307.

EXCEPTION: Handrails in aisles serving seating.

505.4 Height. Top of gripping surfaces of handrails shall be 34 inches (865 mm) minimum and 38 inches (965 mm) maximum vertically above stair nosings, ramp surfaces and walking surfaces. Handrails shall be at a consistent height above stair nosings, ramp surfaces and walking surfaces.

505.5 Clearance. Clearance between handrail gripping surface and adjacent surfaces shall be $1\frac{1}{2}$ inches (38 mm) minimum.

505.6 Gripping Surface. Gripping surfaces shall be continuous, without interruption by newel posts, other construction elements, or obstructions.

EXCEPTIONS:

1. Handrail brackets or balusters attached to the bottom surface of the handrail shall not be considered obstructions, provided they comply with the following criteria:

    a) not more than 20 percent of the handrail length is obstructed,

    b) horizontal projections beyond the sides of the handrail occur $1\frac{1}{2}$ inches (38 mm) minimum below the bottom of the handrail, and provided that for each $\frac{1}{2}$ inch (13 mm) of additional handrail perimeter dimension above 4 inches (100 mm), the vertical clearance dimension of $1\frac{1}{2}$ inch (38 mm) can be reduced by $\frac{1}{8}$ inch (3.2 mm), and

    c) edges shall be rounded.

2. Where handrails are provided along walking surfaces with slopes not steeper than 1:20, the bottoms of handrail gripping surfaces shall be permitted to be obstructed along their entire length where they are integral to crash rails or bumper guards.

505.7 Cross Section. Handrails shall have a cross section complying with Section 505.7.1 or 505.7.2.

505.7.1 Circular Cross Section. Handrails with a circular cross section shall have an outside diameter of $1\frac{1}{4}$ inches (32 mm) minimum and 2 inches (51 mm) maximum.

505.7.2 Noncircular Cross Sections. Handrails with a noncircular cross section shall have a perimeter dimension of 4 inches (100 mm) minimum and $6\frac{1}{4}$ inches (160 mm) maximum, and a cross-section dimension of $2\frac{1}{4}$ inches (57 mm) maximum.

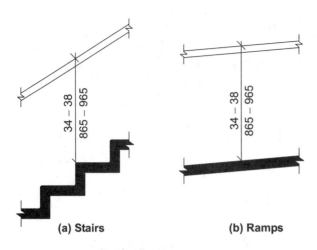

(a) Stairs   (b) Ramps

Fig. 505.4
Handrail Height

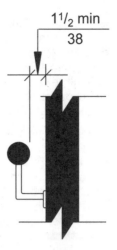

Fig. 505.5
Handrail Clearance

505.8 Surfaces. Handrails, and any wall or other surfaces adjacent to them, shall be free of any sharp or abrasive elements. Edges shall be rounded.

505.9 Fittings. Handrails shall not rotate within their fittings.

505.10 Handrail Extensions. Handrails shall extend beyond and in the same direction of stair flights and ramp runs in accordance with Section 505.10.

EXCEPTIONS:

1. Continuous handrails at the inside turn of stairs and ramps.

2. Extensions are not required for handrails in aisles serving seating where the handrails are discontinuous to provide access to seating and to permit crossovers within the aisle.

3. In alterations, full extensions of handrails shall not be required where such extensions would be hazardous due to plan configuration.

505.10.1 Top and Bottom Extension at Ramps. Ramp handrails shall extend horizontally above the landing 12 inches (305 mm) minimum beyond the top and bottom of ramp runs. Extensions shall return to a wall, guard, or floor, or shall be continuous to the handrail of an adjacent ramp run.

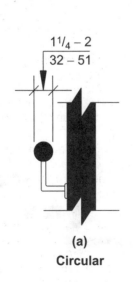

(a)
Circular

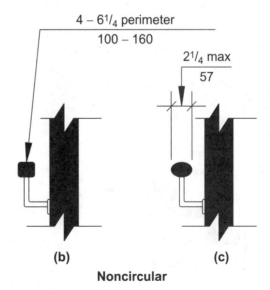

(b)   (c)
Noncircular

Fig. 505.7
Handrail Cross Section

505.10.2 Top Extension at Stairs. At the top of a stair flight, handrails shall extend horizontally above the landing for 12 inches (305 mm) minimum beginning directly above the landing nosing. Extensions shall return to a wall, guard, or the landing surface, or shall be continuous to the handrail of an adjacent stair flight.

505.10.3 Bottom Extension at Stairs. At the bottom of a stair flight, handrails shall extend at the slope of the stair flight for a horizontal distance equal to one tread depth beyond the bottom tread nosing. Extensions shall return to a wall, guard, or the landing surface, or shall be continuous to the handrail of an adjacent stair flight.

506 Windows

Accessible windows shall have operable parts complying with Section 309.

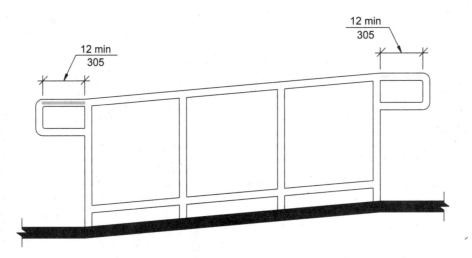

Fig. 505.10.1
Top and Bottom Handrail Extensions at Ramps

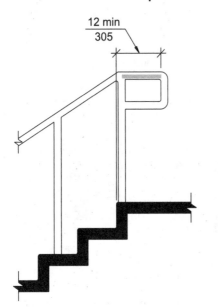

Fig. 505.10.2
Top Handrail Extensions at Stairs

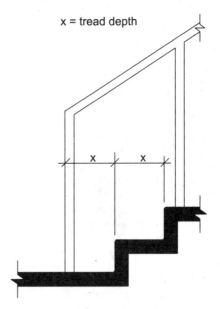

Fig. 505.10.3
Bottom Handrail Extensions at Stairs

Chapter 6. Plumbing Elements and Facilities

601 General

601.1 Scope. Plumbing elements and facilities required to be accessible by scoping provisions adopted by the administrative authority shall comply with the applicable provisions of Chapter 6.

602 Drinking Fountains

602.1 General. Accessible drinking fountains shall comply with Sections 602 and 307.

602.2 Clear Floor Space. A clear floor space complying with Section 305, positioned for a forward approach to the drinking fountain, shall be provided. Knee and toe space complying with Section 306 shall be provided. The clear floor space shall be centered on the drinking fountain.

EXCEPTIONS:

1. Drinking fountains for standing persons.

2. Drinking fountains primarily for children's use shall be permitted where the spout is 30 inches (760 mm) maximum above the floor, and a parallel approach complying with Section 305, centered on the drinking fountain, is provided.

3. In existing buildings, existing drinking fountains providing a parallel approach complying with Section 305, centered on the drinking fountain, shall be permitted.

4. Where specifically permitted by the administrative authority, a parallel approach complying with Section 305, centered on the drinking fountain, shall be permitted for drinking fountains that replace existing drinking fountains with a parallel approach.

602.3 Operable Parts. Operable parts shall comply with Section 309.

602.4 Spout Outlet Height. Spout outlets of wheelchair accessible drinking fountains shall be 36 inches (915 mm) maximum above the floor. Spout outlets of drinking fountains for standing persons shall be 38 inches (965 mm) minimum and 43 inches (1090 mm) maximum above the floor.

602.5 Spout Location. The spout shall be located 15 inches (380 mm) minimum from the vertical support and 5 inches (125 mm) maximum from the front edge of the drinking fountain, including bumpers. Where only a parallel approach is provided, the

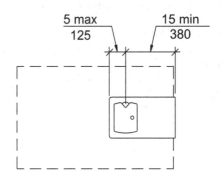

(a) Forward Approach

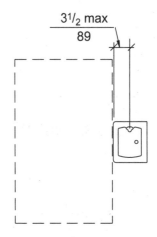

(b) Parallel Approach

Fig. 602.5
Drinking Fountain Spout Location

spout shall be located 3½ inches (89 mm) maximum from the front edge of the drinking fountain, including bumpers.

602.6 Water Flow. The spout shall provide a flow of water 4 inches (102 mm) minimum in height. The angle of the water stream from spouts within 3 inches (76 mm) of the front of the drinking fountain shall be 30 degrees maximum, and from spouts between 3 inches (76 mm) and 5 inches (125 mm) from the front of the drinking fountain shall be 15 degrees maximum, measured horizontally relative to the front face of the drinking fountain.

603 Toilet and Bathing Rooms

603.1 General. Accessible toilet and bathing rooms shall comply with Section 603.

603.2 Clearances.

603.2.1 Turning Space. A turning space complying with Section 304 shall be provided within the room.

603.2.2 Overlap. Clear floor spaces, clearances at fixtures, and turning spaces shall be permitted to overlap.

603.2.3 Door Swing. Doors shall not swing into the clear floor space or clearance for any fixture.

EXCEPTIONS:

1. Doors to a toilet and bathing room for a single occupant, accessed only through a private office and not for common use or public use shall be permitted to swing into the clear floor space, provided the swing of the door can be reversed to meet Section 603.2.3.

2. Where the room is for individual use and a clear floor space complying with Section 305.3 is provided within the room beyond the arc of the door swing.

603.3 Mirrors. Mirrors located above lavatories, sinks or counters shall be mounted with the bottom edge of the reflecting surface 40 inches (1015 mm) maximum above the floor. Mirrors not located above lavatories, sinks or counters shall be mounted with the bottom edge of the reflecting surface 35 inches (890 mm) maximum above the floor.

603.4 Coat Hooks and Shelves. Coat hooks shall be located within one of the reach ranges specified in Section 308. Shelves shall be 40 inches (1015 mm) minimum and 48 inches (1220 mm) maximum above the floor.

604 Water Closets and Toilet Compartments

604.1 General. Accessible water closets and toilet compartments shall comply with Section 604. Compartments containing more than one plumbing fixture shall comply with Section 603. Wheelchair accessible compartments shall comply with Section 604.8. Ambulatory accessible compartments shall comply with Section 604.9.

EXCEPTION: Water closets and toilet compartments primarily for children's use shall be permitted to comply with Section 604.10 as applicable.

604.2 Location. The water closet shall be located with a wall or partition to the rear and to one side. The centerline of the water closet shall be 16 inches (405 mm) minimum to 18 inches (455 mm) maximum from the side wall or partition. Water closets located in ambulatory accessible compartments specified in Section 604.9 shall have the centerline of the water closet 17 inches (430 mm) minimum to 19 inches (485 mm) maximum from the side wall or partition.

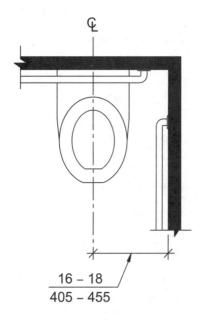

16 – 18
405 – 455

(a) Accessible Water Closets

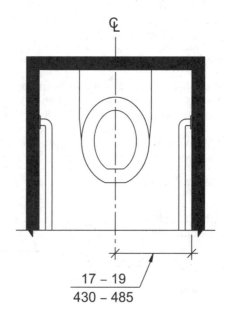

17 – 19
430 – 485

(b) Ambulatory Accessible Water Closets

Fig. 604.2
Water Closet Location

604.3 Clearance.

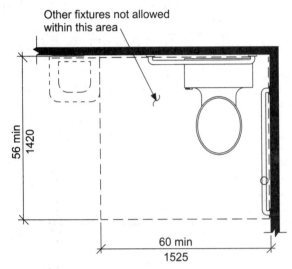

Fig. 604.3
Size of Clearance for Water Closet

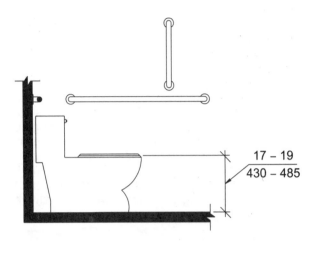

Fig. 604.4
Water Closet Height

604.3.1 Size. A clearance around a water closet 60 inches (1525 mm) minimum, measured perpendicular from the sidewall, and 56 inches (1420 mm) minimum, measured perpendicular from the rear wall, shall be provided.

604.3.2 Overlap. The required clearance around the water closet shall be permitted to overlap the water closet, associated grab bars, paper dispensers, sanitary napkin receptacles, coat hooks, shelves, accessible routes, clear floor space at other fixtures and the turning space. No other fixtures or obstructions shall be within the required water closet clearance.

604.4 Height. The height of water closet seats shall be 17 inches (430 mm) minimum and 19 inches (485 mm) maximum above the floor, measured to the top of the seat. Seats shall not be sprung to return to a lifted position.

> **EXCEPTION:** A water closet in a toilet room for a single occupant, accessed only through a private office and not for common use or public use, shall not be required to comply with Section 604.4.

604.5 Grab Bars. Grab bars for water closets shall comply with Section 609 and shall be provided in accordance with Sections 604.5.1 and 604.5.2. Grab bars shall be provided on the rear wall and on the side wall closest to the water closet.

> **EXCEPTIONS:**
>
> 1. Grab bars are not required to be installed in a toilet room for a single occupant, accessed only through a private office and not for common use or public use, provided reinforcement has been installed in walls

and located so as to permit the installation of grab bars complying with Section 604.5.

> 2. In detention or correction facilities, grab bars are not required to be installed in housing or holding cells or rooms that are specially designed without protrusions for purposes of suicide prevention.
>
> 3. In Type A units, grab bars are not required to be installed where reinforcement complying with Section 1003.11.4 is installed for the future installation of grab bars.
>
> 4. In Type B units located in institutional facilities and assisted living facilities, two swing-up grab bars shall be permitted to be installed in lieu of the rear wall and side wall grab bars. Swing-up grab bars shall comply with Sections 604.5.3 and 609.
>
> 5. In a Type B unit, where fixtures are located on both sides of the water closet, a swing-up grab bar complying with Sections 604.5.3 and 609 shall be permitted. The swing-up grab bar shall be installed on the side of the water closet with the 18 inch (455 mm) clearance required by Section 1004.11.3.1.2.

604.5.1 Fixed Side Wall Grab Bars. Fixed side-wall grab bars shall be 42 inches (1065 mm) minimum in length, located 12 inches (305 mm) maximum from the rear wall and extending 54 inches (1370 mm) minimum from the rear wall. In addition, a vertical grab bar 18 inches (455 mm) minimum in length shall be mounted with the bottom of the bar located between 39 inches (990 mm) and 41 inches (1040 mm) above the floor, and with the center line of the bar located between 39 inches (990 mm) and 41 inches (1040 mm) from the rear wall.

EXCEPTIONS:

1. In Type A and Type B units, the vertical grab bar component is not required.

2. In a Type B unit, when a side wall is not available for a 42-inch (1065 mm) grab bar, the sidewall grab bar shall be permitted to be 18 inches (455 mm) minimum in length, located 12 inches (305 mm) maximum from the rear wall and extending 30 inches (760 mm) minimum from the rear wall.

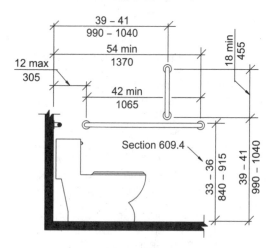

Fig. 604.5.1
Side Wall Grab Bar for Water Closet

604.5.2 Rear Wall Grab Bars. The rear wall grab bar shall be 36 inches (915 mm) minimum in length, and extend from the centerline of the water closet 12 inches (305 mm) minimum on the side closest to the wall, and 24 inches (610 mm) minimum on the transfer side.

EXCEPTIONS:

1. The rear grab bar shall be permitted to be 24 inches (610 mm) minimum in length, centered on the water closet, where wall space does not permit a grab bar 36 inches (915 mm) minimum in length due to the location of a recessed fixture adjacent to the water closet.

2. In a Type A or Type B unit, the rear grab bar shall be permitted to be 24 inches (610 mm) minimum in length, centered on the water closet, where wall space does not permit a grab bar 36 inches (915 mm) minimum in length.

3. Where an administrative authority requires flush controls for flush valves to

be located in a position that conflicts with the location of the rear grab bar, that grab bar shall be permitted to be split or shifted to the open side of the toilet area.

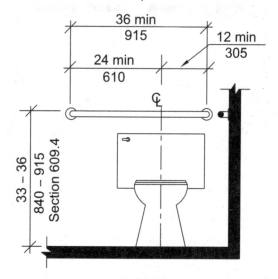

Fig. 604.5.2
Rear Wall Grab Bar for Water Closet

604.5.3 Swing-up Grab Bars. Where swing-up grab bars are installed, a clearance of 18 inches (455 mm) minimum from the centerline of the water closet to any side wall or obstruction shall be provided. A swing-up grab bar shall be installed with the centerline of the grab bar $15^3/_4$ inches (400 mm) from the centerline of the water closet. Swing-up grab bars shall be 28 inches (710 mm) minimum in length, measured from the wall to the end of the horizontal portion of the grab bar.

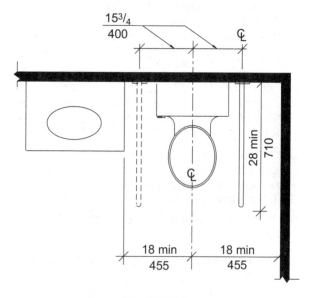

Fig. 604.5.3
Swing-up Grab Bar for Water Closet

604.6 Flush Controls. Flush controls shall be hand operated or automatic. Hand operated flush controls shall comply with Section 309. Flush controls shall be located on the open side of the water closet.

> **EXCEPTION:** In ambulatory accessible compartments complying with Section 604.9, flush controls shall be permitted to be located on either side of the water closet.

604.7 Dispensers. Toilet paper dispensers shall comply with Section 309.4 and shall be 7 inches (180 mm) minimum and 9 inches (230 mm) maximum in front of the water closet measured to the centerline of the dispenser. The outlet of the dispenser shall be 15 inches (380 mm) minimum and 48 inches (1220 mm) maximum above the floor, and shall not be located behind the grab bars. Dispensers shall not be of a type that control delivery, or do not allow continuous paper flow.

604.8 Wheelchair Accessible Compartments.

604.8.1 General. Wheelchair accessible compartments shall comply with Section 604.8.

604.8.2 Size. The minimum area of a wheelchair accessible compartment shall be 60 inches (1525 mm) minimum in width measured perpendicular to the side wall, and 56 inches (1420 mm) minimum in depth for wall hung water closets, and 59 inches (1500 mm) minimum in depth for floor mounted water closets measured perpendicular to the rear wall. The minimum area of a wheelchair accessible compartment for primarily children's use shall be 60 inches (1525 mm) minimum in width measured perpendicular to the side wall, and 59 inches (1500 mm) minimum in

depth for wall hung and floor mounted water closets measured perpendicular to the rear wall.

604.8.3 Doors. Toilet compartment doors, including door hardware, shall comply with Section 404.1, except if the approach is to the latch side of the compartment door clearance between the door side of the stall and any obstruction shall be 42 inches (1065 mm) minimum. Doors shall be located in the front partition or in the side wall or partition farthest from the water closet. Where located in the front partition, the door opening shall be 4 inches (100 mm) maximum from the side wall or partition farthest from the water closet. Where located in the side wall or partition, the door opening shall be 4 inches (100 mm) maximum from the front partition. The door shall be self-closing. A door pull complying with Section 404.2.6 shall be placed on both sides of the door near the latch. Toilet compartment doors shall not swing into the required minimum area of the compartment.

604.8.4 Approach. Wheelchair accessible compartments shall be arranged for left-hand or right-hand approach to the water closet.

604.8.5 Toe Clearance. The front partition and at least one side partition shall provide a toe clearance of 9 inches (230 mm) minimum above the floor and extending 6 inches (150 mm) beyond the compartment side face of the partition, exclusive of partition support members. Compartments primarily for children's use shall provide a toe clearance of 12 inches (305 mm) minimum above the floor and extending 6 inches (150 mm) beyond the compartment side face of

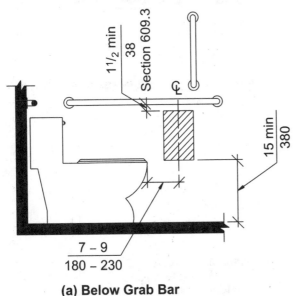

(a) Below Grab Bar   (b) Above Grab Bar

Fig. 604.7
Dispenser Location

the partition, exclusive of partition support members.

EXCEPTIONS:

1. Toe clearance at the front partition is not required in a compartment greater than 62 inches (1575 mm) in depth with a wall-hung water closet, or greater than 65 inches (1650 mm) in depth with a floor-mounted water closet. In a compartment primarily for children's use, greater than 65 inches (1650 mm) in depth, toe clearance at the front partition is not required.

2. Toe clearance at the side partition is not required in a compartment greater than 66 inches (1675 mm) in width.

604.8.6 Grab Bars. Grab bars shall comply with Section 609. Side wall grab bars complying with Section 604.5.1 located on the wall closest to the water closet, and a rear wall grab bar complying with Section 604.5.2, shall be provided.

604.9 Ambulatory Accessible Compartments.

604.9.1 General. Ambulatory accessible compartments shall comply with Section 604.9.

604.9.2 Size. The minimum area of an ambulatory accessible compartment shall be 60 inches (1525 mm) minimum in depth and 36 inches (915 mm) in width.

604.9.3 Doors. Toilet compartment doors, including door hardware, shall comply with Sec-

tion 404, except if the approach is to the latch side of the compartment door the clearance between the door side of the compartment and any obstruction shall be 42 inches (1065 mm) minimum. The door shall be self-closing. A door pull complying with Section 404.2.6 shall be placed on both sides of the door near the latch. Compartment doors shall not swing into the required minimum area of the compartment.

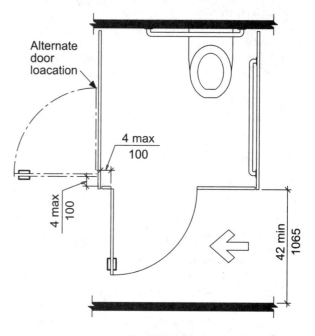

Fig. 604.8.3
Wheelchair Accessible Compartment Doors

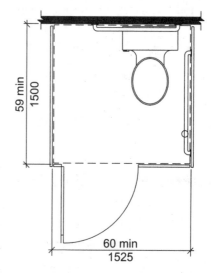

(a) Wall-Hung Water Closet – Adult

(b) Floor-Mounted Water Closet – Adult
Wall-Hung and
Floor-Mounted Water Closet – Children

Fig. 604.8.2
Wheelchair Accessible Toilet Compartments

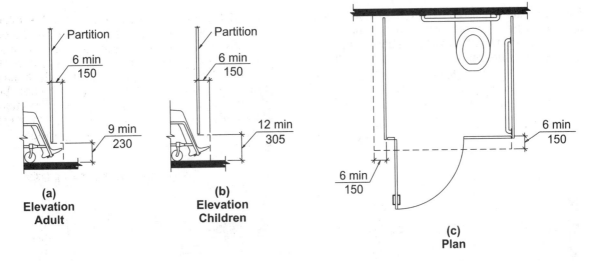

(a)
Elevation
Adult

(b)
Elevation
Children

(c)
Plan

Fig. 604.8.5
Wheelchair Accessible Compartment Toe Clearance

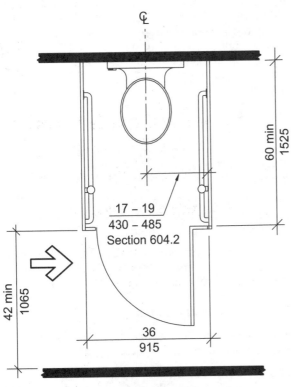

Fig. 604.9
Ambulatory Accessible Compartment

604.9.4 Grab Bars. Grab bars shall comply with Section 609. Side wall grab bars complying with Section 604.5.1 shall be provided on both sides of the compartment.

604.10 Water Closets and Toilet Compartments for Children's Use.

604.10.1 General. Accessible water closets and toilet compartments primarily for children's use shall comply with Section 604.10.

604.10.2 Location. The water closet shall be located with a wall or partition to the rear and to one side. The centerline of the water closet shall be 12 inches (305 mm) minimum to 18 inches (455 mm) maximum from the side wall or partition. Water closets located in ambulatory accessible toilet compartments specified in Section 604.9 shall be located as specified in Section 604.2.

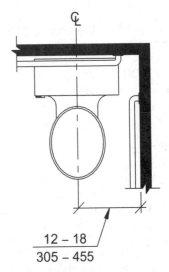

Fig. 604.10.2
Children's Water Closet Location

604.10.3 Clearance. A clearance around a water closet complying with Section 604.3 shall be provided.

604.10.4 Height. The height of water closet seats shall be 11 inches (280 mm) minimum and 17 inches (430 mm) maximum above the floor, measured to the top of the seat. Seats shall not be sprung to return to a lifted position.

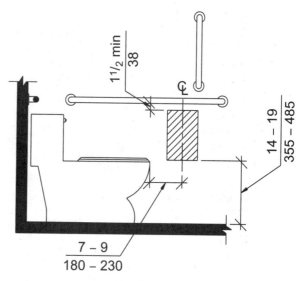

Fig. 604.10.7
Children's Dispenser Location

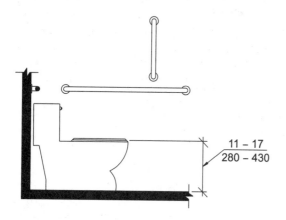

Fig. 604.10.4
Children's Water Closet Height

604.10.5 Grab Bars. Grab bars for water closets shall comply with Section 604.5.

604.10.6 Flush Controls. Flush controls shall be hand operated or automatic. Hand operated flush controls shall comply with Sections 309.2 and 309.4 and shall be installed 36 inches (915 mm) maximum above the floor. Flush controls shall be located on the open side of the water closet.

> **EXCEPTION:** In ambulatory accessible compartments complying with Section 604.9, flush controls shall be permitted to be located on either side of the water closet.

604.10.7 Dispensers. Toilet paper dispensers shall comply with Section 309.4 and shall be 7 inches (180 mm) minimum and 9 inches (230 mm) maximum in front of the water closet measured to the center line of the dispenser. The outlet of the dispenser shall be 14 inches (355 mm) minimum and 19 inches (485 mm) maximum above the floor. There shall be a clearance of $1\frac{1}{2}$ inches (38 mm) minimum below the grab bar. Dispensers shall not be of a type that control delivery or do not allow continuous paper flow.

604.10.8 Toilet Compartments. Toilet compartments shall comply with Sections 604.8 and 604.9, as applicable.

604.11 Coat Hooks and Shelves. Coat hooks provided within toilet compartments shall be 48 inches (1220 mm) maximum above the floor. Shelves shall be 40 inches (1015 mm) minimum and 48 inches (1220 mm) maximum above the floor.

605 Urinals

605.1 General. Accessible urinals shall comply with Section 605.

605.2 Height. Urinals shall be of the stall type or shall be of the wall hung type with the rim at 17 inches (430 mm) maximum above the floor.

605.3 Clear Floor Space. A clear floor space complying with Section 305, positioned for forward approach, shall be provided.

605.4 Flush Controls. Flush controls shall be hand operated or automatic. Hand operated flush controls shall comply with Section 309.

606 Lavatories and Sinks

606.1 General. Accessible lavatories and sinks shall comply with Section 606.

606.2 Clear Floor Space. A clear floor space complying with Section 305.3, positioned for forward approach, shall be provided. Knee and toe clearance complying with Section 306 shall be provided. The dip of the overflow shall not be considered in determining knee and toe clearances.

> **EXCEPTIONS:**
>
> 1. A parallel approach complying with Section 305 shall be permitted to a kitchen sink in a space where a cook top or conventional range is not provided.

2. The requirement for knee and toe clearance shall not apply to a lavatory in a toilet and bathing facility for a single occupant, accessed only through a private office and not for common use or public use.

3. A knee clearance of 24 inches (610 mm) minimum above the floor shall be permitted at lavatories and sinks used primarily by children ages 6 through 12 where the rim or counter surface is 31 inches (785 mm) maximum above the floor.

4. A parallel approach complying with Section 305 shall be permitted at lavatories and sinks used primarily by children ages 5 and younger.

5. The requirement for knee and toe clearance shall not apply to more than one bowl of a multibowl sink.

6. A parallel approach shall be permitted at wet bars.

606.3 Height. The front of lavatories and sinks shall be 34 inches (865 mm) maximum above the floor, measured to the higher of the rim or counter surface.

EXCEPTION: A lavatory in a toilet and bathing facility for a single occupant, accessed only through a private office and not for common use or public use, shall not be required to comply with Section 606.3.

606.4 Faucets. Faucets shall comply with Section 309. Hand-operated metering faucets shall remain open for 10 seconds minimum.

606.5 Lavatories with Enhanced Reach Range. Where enhanced reach range is required at lavatories, faucets and soap dispenser controls shall have a reach depth of 11 inches (280 mm) maximum or, if automatic, shall be activated within a reach depth of 11 inches (280 mm) maximum. Water and soap flow shall be provided with a reach depth of 11 inches (280 mm) maximum.

EXCEPTION: In Type A and Type B units, reach range for lavatory faucets and soap dispensers is not required.

606.6 Exposed Pipes and Surfaces. Water supply and drainpipes under lavatories and sinks shall be insulated or otherwise configured to protect against contact. There shall be no sharp or abrasive surfaces under lavatories and sinks.

606.7 Operable Parts. Operable parts on towel dispensers and hand dryers shall comply with Table 606.7.

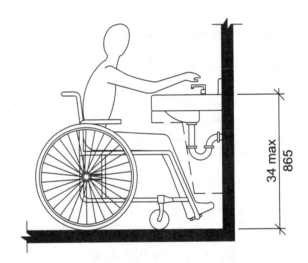

Fig. 606.3
Height of Lavatories and Sinks

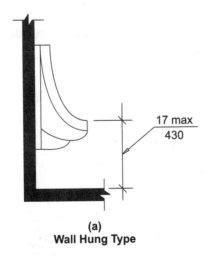

(a)
Wall Hung Type

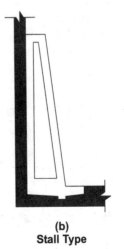

(b)
Stall Type

Fig. 605.2
Height of Urinals

Table 606.7
Maximum Reach Depth and Height

Maximum Reach Depth	5 inch (13 mm)	2 inches (50 mm)	5 inches (125 mm)	6 inches (150 mm)	9 inches (230 mm)	11 inches (280 mm)
Maximum Reach Height	48 inch (1220 mm)	46 inch (1170 mm)	42 inches (1065 mm)	40 inches (1015 mm)	36 inches (915 mm)	34 inches (865 mm)

607 Bathtubs

607.1 General. Accessible bathtubs shall comply with Section 607.

607.2 Clearance. A clearance in front of bathtubs extending the length of the bathtub and 30 inches (760 mm) minimum in depth shall be provided. Where a permanent seat is provided at the head end of the bathtub, the clearance shall extend 12 inches (305 mm) minimum beyond the wall at the head end of the bathtub.

607.3 Seat. A permanent seat at the head end of the bathtub or a removable in-tub seat shall be provided. Seats shall comply with Section 610.

607.4 Grab Bars. Grab bars shall comply with Section 609 and shall be provided in accordance with Section 607.4.1 or 607.4.2.

EXCEPTIONS:

1. Grab bars shall not be required to be installed in a bathing facility for a single occupant accessed only through a private office and not for common use or public use, provided reinforcement has been installed in walls and located so as to permit the installation of grab bars complying with Section 607.4.

2. In Type A units, grab bars are not required to be installed where reinforcement complying with Section 1003.11.9 is installed for the future installation of grab bars.

607.4.1 Bathtubs with Permanent Seats. For bathtubs with permanent seats, grab bars complying with Section 607.4.1 shall be provided.

607.4.1.1 Back Wall. Two horizontal grab bars shall be provided on the back wall, one complying with Section 609.4 and the other 9 inches (230 mm) above the rim of the bathtub. Each grab bar shall be located 15 inches (380 mm) maximum from the head end wall and extend to 12 inches (305 mm) maximum from the control end wall.

607.4.1.2 Control End Wall. Control end wall grab bars shall comply with Section 607.4.1.2.

EXCEPTION: An L-shaped continuous grab bar of equivalent dimensions and positioning shall be permitted to serve the function of separate vertical and horizontal grab bars.

607.4.1.2.1 Horizontal Grab Bar. A horizontal grab bar 24 inches (610 mm) mini-

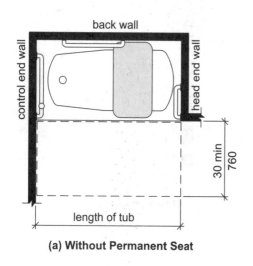

(a) Without Permanent Seat

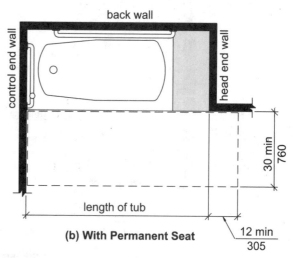

(b) With Permanent Seat

Fig. 607.2
Clearance for Bathtubs

mum in length shall be provided on the control end wall at the front edge of the bathtub.

607.4.1.2.2 Vertical Grab Bar. A vertical grab bar 18 inches (455 mm) minimum in length shall be provided on the control end wall 3 inches (75 mm) minimum to 6 inches (150 mm) maximum above the horizontal grab bar, and 4 inches (100 mm) maximum inward from the front edge of the bathtub.

607.4.2 Bathtubs without Permanent Seats. For bathtubs without permanent seats, grab bars complying with Section 607.4.2 shall be provided.

607.4.2.1 Back Wall. Two horizontal grab bars shall be provided on the back wall, one complying with Section 609.4 and the other 9 inches (230 mm) above the rim of the bathtub. Each grab bar shall be 24 inches (610 mm) minimum in length, located 24 inches (610 mm) maximum from the head end wall and extend to 12 inches (305 mm) maximum from the control end wall.

607.4.2.2 Control End Wall. Control end wall grab bars shall comply with Section 607.4.2.2.

EXCEPTION: An L-shaped continuous grab bar of equivalent dimensions and positioning shall be permitted to serve the function of separate vertical and horizontal grab bars.

607.4.2.2.1 Horizontal Grab Bar. A horizontal grab bar 24 inches (610 mm) minimum in length shall be provided on the control end wall beginning near the front edge of the bathtub and extend toward the inside corner of the bathtub.

607.4.2.2.2 Vertical Grab Bar. A vertical grab bar 18 inches (455 mm) minimum in length shall be provided on the control end wall 3 inches (76 mm) minimum to 6 inches (150 mm) maximum above the horizontal grab bar, and 4 inches (102 mm) maximum inward from the front edge of the bathtub.

607.4.2.3 Head End Wall. A horizontal grab bar 12 inches (305 mm) minimum in length shall be provided on the head end wall at the front edge of the bathtub.

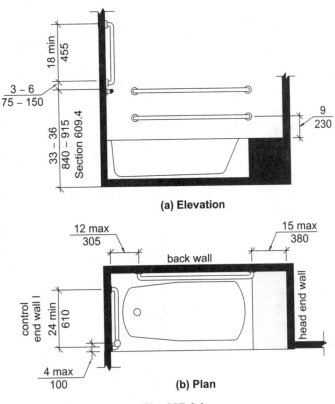

Fig. 607.4.1
Grab Bars for Bathtubs with Permanent Seats

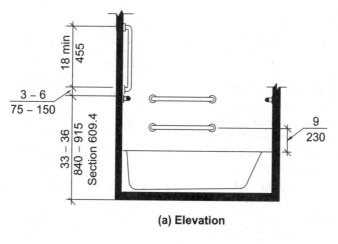

(a) Elevation

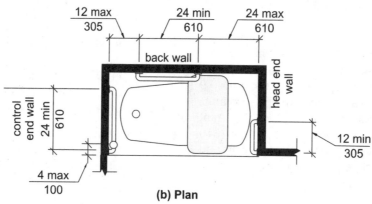

(b) Plan

Fig. 607.4.2
Grab Bars for Bathtubs without Permanent Seats

607.5 Controls. Controls, other than drain stoppers, shall be provided on an end wall, located between the bathtub rim and grab bar, and between the open side of the bathtub and the midpoint of the width of the bathtub. Controls shall comply with Section 309.4.

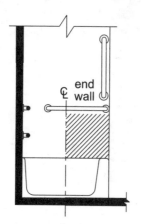

Fig. 607.5
Location of Bathtub Controls

607.6 Hand Shower. A hand shower with a hose 59 inches (1500 mm) minimum in length, that can be used as both a fixed shower head and as a hand shower, shall be provided. The hand shower shall have a control with a nonpositive shut-off feature. An adjustable-height hand shower mounted on a vertical bar shall be installed so as to not obstruct the use of grab bars.

607.7 Bathtub Enclosures. Enclosures for bathtubs shall not obstruct controls or transfer from wheelchairs onto bathtub seats or into bathtubs. Enclosures on bathtubs shall not have tracks installed on the rim of the bathtub.

607.8 Water Temperature. Bathtubs shall deliver water that is 120 degrees F (49 degrees C) maximum.

608 Shower Compartments

608.1 General. Accessible shower compartments shall comply with Section 608.

608.2 Size and Clearances.

608.2.1 Transfer-Type Shower Compartments.
Transfer-type shower compartments shall have a clear inside dimension of 36 inches (915 mm) in width and 36 inches (915 mm) in depth, measured at the center point of opposing sides. An entry 36 inches (915 mm) minimum in width shall be provided. A clearance of 48 inches (1220 mm) minimum in length measured perpendicular from the control wall, and 36 inches (915 mm) minimum in depth shall be provided adjacent to the open face of the compartment.

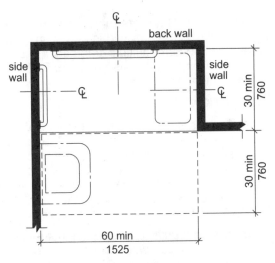

Note: inside finished dimensions measured at the center points of opposing sides

Fig. 608.2.2
Standard Roll-in-Type Shower Compartment
Size and Clearance

608.2.3 Alternate Roll-in-Type Shower Compartments.
Alternate roll-in shower compartments shall have a clear inside dimension of 60 inches (1525 mm) minimum in width, and 36 inches (915 mm) in depth, measured at the center point of opposing sides. An entry 36 inches (915) mm) minimum in width shall be provided at one end of the 60-inch (1525 mm) width of the compartment. A seat wall, 24 inches (610 mm) minimum and 36 inches (915 mm) maximum in length, shall be provided on the entry side of the compartment.

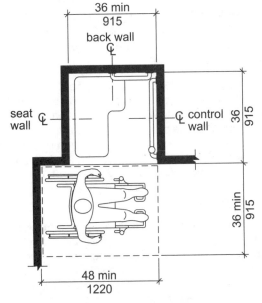

Note: inside finished dimensions measured at the center points of opposing sides

Fig. 608.2.1
Transfer-Type Shower Compartment
Size and Clearance

608.2.2 Standard Roll-in-Type Shower Compartments.
Standard roll-in-type shower compartments shall have a clear inside dimension of 60 inches (1525 mm) minimum in width and 30 inches (760 mm) minimum in depth, measured at the center point of opposing sides. An entry 60 inches (1525 mm) minimum in width shall be provided. A clearance of 60 inches (1525 mm) minimum in length adjacent to the 60-inch (1525 mm) width of the open face of the shower compartment, and 30 inches (760 mm) minimum in depth, shall be provided. A lavatory complying with Section 606 shall be permitted at the end of the clearance opposite the shower compartment side where shower controls are positioned. Where shower controls are located on the back wall and no seat is provided, the lavatory shall be permitted at either end of the clearance.

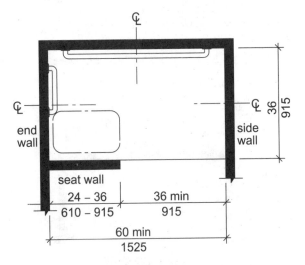

Note: inside finished dimensions measured at the center points of opposing sides

Fig. 608.2.3
Alternate Roll-in-Type Shower Compartment
Size and Clearance

608.3 Grab Bars. Grab bars shall comply with Section 609 and shall be provided in accordance with Section 608.3. Where multiple grab bars are used, required horizontal grab bars shall be installed at the same height above the floor.

EXCEPTIONS:

1. Grab bars are not required to be installed in a shower facility for a single occupant, accessed only through a private office and not for common use or public use, provided reinforcement has been installed in walls and located so as to permit the installation of grab bars complying with Section 608.3.

2. In Type A units, grab bars are not required to be installed where reinforcement complying with Section 1003.11.9 is installed for the future installation of grab bars.

608.3.1 Transfer-Type Showers. Grab bars for transfer type showers shall comply with Section 608.3.1.

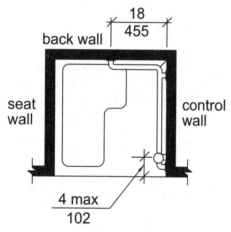

Fig. 608.3.1
Grab Bars in Transfer-Type Showers

608.3.1.1 Horizontal Grab Bars. Horizontal grab bars shall be provided across the control wall and on the back wall to a point 18 inches (455 mm) from the control wall.

608.3.1.2 Vertical Grab Bar. A vertical grab bar 18 inches (455 mm) minimum in length shall be provided on the control end wall 3 inches (75 mm) minimum to 6 inches (150 mm) maximum above the horizontal grab bar, and 4 inches (100 mm) maximum inward from the front edge of the shower.

608.3.1.3 Grab Bar Configuration. Grab bars complying with Sections 608.3.1.1 and 608.3.1.2 shall be permitted to be separate bars, a single piece bar, or combination thereof.

608.3.2 Standard Roll-in-Type Showers. In standard roll-in type showers, grab bars shall be provided on three walls of showers without seats. Where a seat is provided in a standard roll-in type shower, grab bars shall be provided on the back wall and on the wall opposite the seat. Grab bars shall not be provided above the seat. Grab bars shall be 6 inches (150 mm) maximum from the adjacent wall.

608.3.3 Alternate Roll-in-Type Showers. In alternate roll-in type showers, grab bars shall be provided on the back wall and the end wall adjacent to the seat. Grab bars shall not be provided above the seat. Grab bars shall be 6 inches (150 mm) maximum from the adjacent wall.

608.4 Seats. A folding or nonfolding seat shall be provided in transfer-type shower compartments. A seat shall be provided in an alternate roll-in-type shower compartment. In standard and alternate

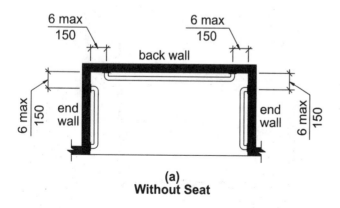

(a)
Without Seat

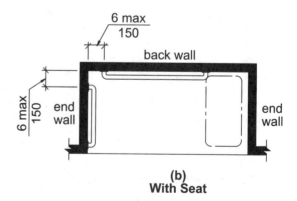

(b)
With Seat

Fig. 608.3.2
Grab Bars in Standard Roll-in-Type Showers

roll-in-type showers where a seat is provided, if the seat extends over the minimum clear inside dimension required by Section 608.2.2 or 608.2.3, the seat shall be a folding seat. Seats shall comply with Section 610.

EXCEPTIONS:

1. A shower seat is not required to be installed in a shower facility for a single occupant, accessed only through a private office and not for common use or public use, provided reinforcement has been installed in walls and located so as to permit the installation of a shower seat complying with Section 608.4.

2. In Type A units, a shower seat is not required to be installed where reinforcement complying with Section 1003.11.9 is installed for the future installation of a shower seat.

608.5 Controls and Hand Showers. Controls and hand showers shall comply with Sections 608.5 and 309.4.

608.5.1 Transfer-Type Showers. In transfer-type showers, the controls and hand shower shall be located on the control wall opposite the seat, 38 inches (965 mm) minimum and 48 inches (1220 mm) maximum above the shower floor, within 15 inches (380 mm), left or right, of the centerline of the seat.

608.5.2 Standard Roll-in Showers. In standard roll-in showers, the controls and hand shower shall be located 38 inches (965 mm) minimum and 48 inches (1220 mm) maximum above the shower floor. In standard roll-in showers with seats, the controls and hand shower shall be located on the back wall, no more than 27 inches (685 mm) maximum from the end wall behind the seat.

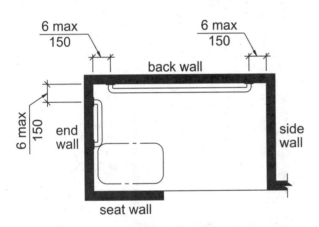

Fig. 608.3.3
Grab Bars in Alternate Roll-in-Type Showers

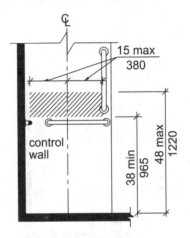

Fig. 608.5.1
Transfer-Type Shower
Controls and Handshowers Location

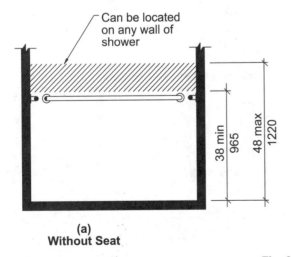

(a)
Without Seat

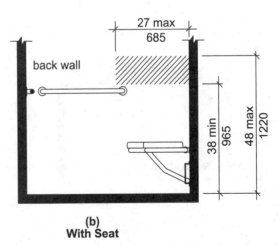

(b)
With Seat

Fig. 608.5.2
Standard Roll-in-Type Shower Control and Handshower Location

608.5.3 Alternate Roll-in Showers. In alternate roll-in showers, the controls and hand shower shall be located 38 inches (965 mm) minimum and 48 inches (1220 mm) maximum above the shower floor. In alternate roll-in showers with controls and hand shower located on the end wall adjacent to the seat, the controls and hand shower shall be 27 inches (685 mm) maximum from the seat wall. In alternate roll-in showers with the controls and hand shower located on the back wall opposite the seat, the controls and hand shower shall be located within 15 inches (380 mm), left or right, of the centerline of the seat.

EXCEPTION: A fixed shower head with the controls and shower head located on the back wall opposite the seat shall be permitted.

608.6 Hand Showers. A hand shower with a hose 59 inches (1500 mm) minimum in length, that can be used both as a fixed shower head and as a hand shower, shall be provided. The hand shower shall have a control with a nonpositive shut-off feature. An adjustable-height shower head mounted on a vertical bar shall be installed so as to not obstruct the use of grab bars.

EXCEPTION: A fixed shower head shall be permitted in lieu of a hand shower where the scoping provisions of the administrative authority require a fixed shower head.

608.7 Thresholds. Thresholds in roll-in-type shower compartment shall be $1/2$ inch (13 mm) maximum in height in accordance with Section 303. In transfer-type shower compartments, thresholds $1/2$ inch (13 mm) maximum in height shall be beveled, rounded, or vertical.

EXCEPTION: In existing facilities, in transfer-type shower compartments where provision of a threshold $1/2$ inch (13 mm) in height would disturb the structural reinforcement of the floor slab, a threshold 2 inches (51 mm) maximum in height shall be permitted.

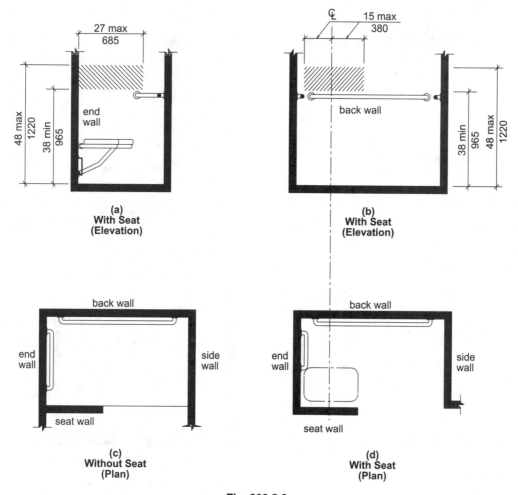

Fig. 608.5.3
Alternate Roll-in-Type Shower Control and Handshower Location

608.8 Shower Enclosures. Shower compartment enclosures for shower compartments shall not obstruct controls or obstruct transfer from wheelchairs onto shower seats.

608.9 Water Temperature. Showers shall deliver water that is 120 degrees F (49 degrees C) maximum.

609 Grab Bars

609.1 General. Grab bars in accessible toilet or bathing facilities shall comply with Section 609.

609.2 Cross Section. Grab bars shall have a cross section complying with Section 609.2.1 or 609.2.2.

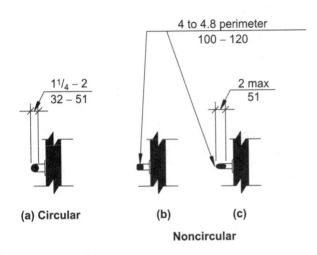

Fig. 609.2
Size of Grab Bars

609.2.1 Circular Cross Section. Grab bars with a circular cross section shall have an outside diameter of $1\frac{1}{4}$ inch (32 mm) minimum and 2 inches (51 mm) maximum.

609.2.2 Noncircular Cross Section. Grab bars with a noncircular cross section shall have a cross section dimension of 2 inches (51 mm) maximum, and a perimeter dimension of 4 inches (102 mm) minimum and 4.8 inches (122 mm) maximum.

609.3 Spacing. The space between the wall and the grab bar shall be $1\frac{1}{2}$ inches (38 mm). The space between the grab bar and projecting objects below and at the ends of the grab bar shall be $1\frac{1}{2}$ inches (38 mm) minimum. The space between the grab bar and projecting objects above the grab bar shall be 12 inches (305 mm) minimum.

EXCEPTIONS:

　　1. The space between the grab bars and shower controls, shower fittings, and other

grab bars above the grab bar shall be permitted to be $1\frac{1}{2}$ inches (38 mm) minimum.

　　2. Swing-up grab bars shall not be required to comply with Section 609.3.

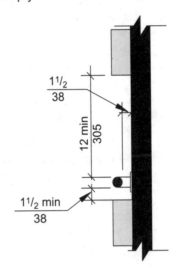

Fig. 609.3
Spacing of Grab Bars

609.4 Position of Grab Bars. Grab bars shall be installed in a horizontal position, 33 inches (840 mm) minimum and 36 inches (915 mm) maximum above the floor measured to the top of the gripping surface. At water closets primarily for children's use complying with Section 604.10, grab bars shall be installed in a horizontal position 18 inches (455 mm) minimum to 27 inches (685 mm) maximum above the floor measured to the top of the gripping surface.

EXCEPTIONS:

　　1. The lower grab bar on the back wall of a bathtub required by Section 607.4.1.1 or 607.4.2.1.

　　2. Vertical grab bars required by Sections 604.5.1, 607.4.1.2.2, 607.4.2.2.2, and 608.3.1.2.

609.5 Surface Hazards. Grab bars, and any wall or other surfaces adjacent to grab bars, shall be free of sharp or abrasive elements. Edges shall be rounded.

609.6 Fittings. Grab bars shall not rotate within their fittings.

609.7 Installation. Grab bars shall be installed in any manner that provides a gripping surface at the locations specified in this standard and does not obstruct the clear floor space.

609.8 Structural Strength. Allowable stresses shall not be exceeded for materials used where a vertical or horizontal force of 250 pounds (1112 N) is applied at any point on the grab bar, fastener mounting device, or supporting structure.

610 Seats

610.1 General. Seats in accessible bathtubs and shower compartments shall comply with Section 610.

610.2 Bathtub Seats. The height of bathtub seats shall be 17 inches (430 mm) minimum to 19 inches (485 mm) maximum above the bathroom floor, measured to the top of the seat. Removable in-tub seats shall be 15 inches (380 mm) minimum and 16 inches (405 mm) maximum in depth. Removable in-tub seats shall be capable of secure placement. Permanent seats shall be 15 inches (380 mm) minimum in depth and shall extend from the back wall to or beyond the outer edge of the bathtub. Permanent seats shall be positioned at the head end of the bathtub.

610.3 Shower Compartment Seats. Where a seat is provided in a standard roll-in shower compartment, it shall be a folding type and shall be on the wall adjacent to the controls. The height of the seat shall be 17 inches (430 mm) minimum and 19 inches (485 mm) maximum above the bathroom floor, measured to the top of the seat. In trans-

fer-type and alternate roll-in-type showers, the seat shall extend along the seat wall to a point within 3 inches (75 mm) of the compartment entry. In standard roll-in-type showers, the seat shall extend from the control wall to a point within 3 inches (75 mm) of the compartment entry. Seats shall comply with Section 610.3.1 or 610.3.2.

610.3.1 Rectangular Seats. The rear edge of a rectangular seat shall be 2$\frac{1}{2}$ inches (64 mm) maximum and the front edge 15 inches (380 mm) minimum to 16 inches (405 mm) maximum from the seat wall. The side edge of the seat shall be 1$\frac{1}{2}$ inches (38 mm) maximum from the back wall of a transfer-type shower and 1$\frac{1}{2}$ inches (38 mm) maximum from the control wall of a roll-in-type shower.

610.3.2 L-Shaped Seats. The rear edge of an L-shaped seat shall be 2$\frac{1}{2}$ inches (64 mm) maximum and the front edge 15 inches (380 mm) minimum to 16 inches (405 mm) maximum from the seat wall. The rear edge of the "L" portion of the seat shall be 1$\frac{1}{2}$ inches (38 mm) maximum from the wall and the front edge shall be 14 inches (355 mm) minimum and 15 inches (380 mm) maximum from the wall. The end of the "L" shall be 22 inches (560 mm) minimum and 23 inches (585 mm) maximum from the main seat wall.

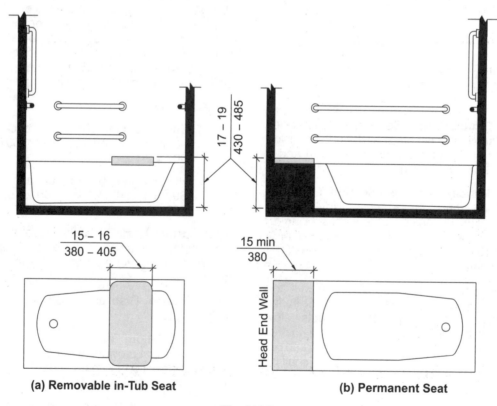

(a) Removable in-Tub Seat

(b) Permanent Seat

**Fig. 610.2
Bathtub Seats**

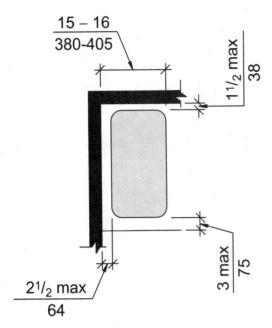

Fig. 610.3.1
Rectangular Shower Compartment Seat

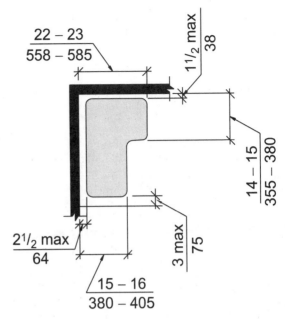

Fig. 610.3.2
L-Shaped Shower Compartment Seat

610.4 Structural Strength. Allowable stresses shall not be exceeded for materials used where a vertical or horizontal force of 250 pounds (1112 N) is applied at any point on the seat, fastener mounting device, or supporting structure.

611 Washing Machines and Clothes Dryers

611.1 General. Accessible washing machines and clothes dryers shall comply with Section 611.

611.2 Clear Floor Space. A clear floor space complying with Section 305, positioned for parallel approach, shall be provided. The clear floor space shall be centered on the appliance.

611.3 Operable Parts. Operable parts, including doors, lint screens, detergent and bleach compartments, shall comply with Section 309.

611.4 Height. Top loading machines shall have the door to the laundry compartment 36 inches (915 mm) maximum above the floor. Front loading machines shall have the bottom of the opening to the laundry compartment 15 inches (380 mm) minimum and 34 inches (865 mm) maximum above the floor.

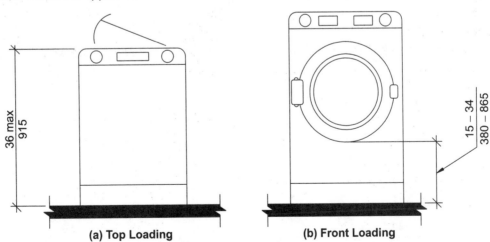

Fig. 611.4
Height of Laundry Equipment

Chapter 7. Communication Elements and Features

701 General

701.1 Scope. Communications elements and features required to be accessible by the scoping provisions adopted by the administrative authority shall comply with the applicable provisions of Chapter 7.

702 Alarms

702.1 General. Accessible audible and visual alarms and notification appliances shall be installed in accordance with NFPA 72 listed in Section 105.2.2, be powered by a commercial light and power source, be permanently connected to the wiring of the premises electric system, and be permanently installed.

703 Signs

703.1 General. Accessible signs shall comply with Section 703.

703.2 Visual Characters.

703.2.1 General. Visual characters shall comply with Section 703.2.

EXCEPTION: Visual characters complying with Section 703.3 shall not be required to comply with Section 703.2.

703.2.2 Case. Characters shall be uppercase, lowercase, or a combination of both.

703.2.3 Style. Characters shall be conventional in form. Characters shall not be italic, oblique, script, highly decorative, or of other unusual forms.

703.2.4 Character Height. The uppercase letter "I" shall be used to determine the allowable height of all characters of a font. The uppercase letter "I" of the font shall have a minimum height complying with Table 703.2.4. Viewing distance shall be measured as the horizontal distance between the character and an obstruction preventing further approach towards the sign.

703.2.5 Character Width. The uppercase letter "O" shall be used to determine the allowable width of all characters of a font. The width of the uppercase letter "O" of the font shall be 55 percent minimum and 110 percent maximum of the height of the uppercase "I" of the font.

703.2.6 Stroke Width. The uppercase letter "I" shall be used to determine the allowable stroke width of all characters of a font. The stroke width shall be 10 percent minimum and 30 percent maximum of the height of the uppercase "I" of the font.

703.2.7 Character Spacing. Spacing shall be measured between the two closest points of adjacent characters within a message, excluding word spaces. Spacing between individual characters shall be 10 percent minimum and 35 percent maximum of the character height.

Table 703.2.4—Visual Character Height

Height above Floor to Baseline of Character	Horizontal Viewing Distance	Minimum Character Height
40 inches (1015 mm) to less than or equal to 70 inches (1780 mm)	Less than 6 feet (1830 mm)	$^5/_8$ inch (16 mm)
	6 feet (1830 mm) and greater	$^5/_8$ inch (16 mm), plus $^1/_8$ inch (3.2 mm) per foot (305 mm) of viewing distance above 6 feet (1830 mm)
Greater than 70 inches (1780 mm) to less than or equal to 120 inches (3050 mm)	Less than 15 feet (4570 mm)	2 inches (51 mm)
	15 feet (4570 mm) and greater	2 inches (51 mm), plus $^1/_8$ inch (3.2 mm) per foot (305 mm) of viewing distance above 15 feet (4570 mm)
Greater than 120 inches (3050 mm)	Less than 21 feet (6400 mm)	3 inches (75 mm)
	21 feet (6400 mm) and greater	3 inches (76 mm), plus $^1/_8$ inch (3.2 mm) per foot (305 mm) of viewing distance above 21 feet (6400 mm)

703.2.8 Line Spacing. Spacing between the baselines of separate lines of characters within a message shall be 135 percent minimum to 170 percent maximum of the character height.

703.2.9 Height Above Floor. Visual characters shall be 40 inches (1015 mm) minimum above the floor of the viewing position, measured to the baseline of the character. Heights shall comply with Table 703.2.4, based on the size of the characters on the sign.

> **EXCEPTION:** Visual characters indicating elevator car controls shall not be required to comply with Section 703.2.9.

703.2.10 Finish and Contrast. Characters and their background shall have a nonglare finish. Characters shall contrast with their background, with either light characters on a dark background, or dark characters on a light background.

703.3 Tactile Characters.

703.3.1 General. Tactile characters shall comply with Section 703.3, and shall be duplicated in braille complying with Section 703.4.

703.3.2 Depth. Tactile characters shall be raised $^1/_{32}$ inch (0.8 mm) minimum above their background.

703.3.3 Case. Characters shall be uppercase.

703.3.4 Style. Characters shall be sans serif. Characters shall not be italic, oblique, script, highly decorative, or of other unusual forms.

703.3.5 Character Height. The uppercase letter "I" shall be used to determine the allowable height of all characters of a font. The height of the uppercase letter "I" of the font, measured vertically from the baseline of the character, shall be $^5/_8$ inch (16 mm) minimum, and 2 inches (51 mm) maximum.

> **EXCEPTION:** Where separate tactile and visual characters with the same information are provided, the height of the tactile uppercase letter "I" shall be permitted to be $^1/_2$ inch (13 mm) minimum.

703.3.6 Character Width. The uppercase letter "O" shall be used to determine the allowable width of all characters of a font. The width of the uppercase letter "O" of the font shall be 55 percent minimum and 110 percent maximum of the height of the uppercase "I" of the font.

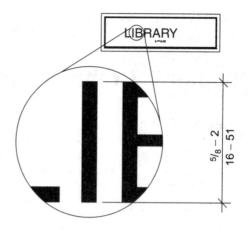

Fig. 703.3.5
Character Height

703.3.7 Stroke Width. Tactile character stroke width shall comply with Section 703.3.7. The uppercase letter "I" of the font shall be used to determine the allowable stroke width of all characters of a font.

703.3.7.1 Maximum. The stroke width shall be 15 percent maximum of the height of the uppercase letter "I" measured at the top surface of the character, and 30 percent maximum of the height of the uppercase letter "I" measured at the base of the character.

703.3.7.2 Minimum. When characters are both visual and tactile, the stroke width shall be 10 percent minimum of the height of the uppercase letter "I".

703.3.8 Character Spacing. Character spacing shall be measured between the two closest points of adjacent tactile characters within a message, excluding word spaces. Spacing between individual tactile character shall be $^1/_8$ inch (3.2 mm) minimum measured at the top surface of the characters, $^1/_{16}$ inch (1.6 mm) minimum measured at the base of the characters, and four times the tactile character stroke width maximum. Characters shall be separated from raised borders and decorative elements $^3/_8$ inch (9.5 mm) minimum.

703.3.9 Line Spacing. Spacing between the baselines of separate lines of tactile characters within a message shall be 135 percent minimum and 170 percent maximum of the tactile character height.

703.3.10 Height above Floor. Tactile characters shall be 48 inches (1220 mm) minimum above the floor, measured to the baseline of the lowest tactile character and 60 inches (1525 mm) maximum above the floor, measured to the baseline of the highest tactile character.

EXCEPTION: Tactile characters for elevator car controls shall not be required to comply with Section 703.3.10.

703.3.11 Location. Where a tactile sign is provided at a door, the sign shall be alongside the door at the latch side. Where a tactile sign is provided at double doors with one active leaf, the sign shall be located on the inactive leaf. Where a tactile sign is provided at double doors with two active leaves, the sign shall be to the right of the right-hand door. Where there is no wall space on the latch side of a single door, or to the right side of double doors, signs shall be on the nearest adjacent wall. Signs containing tactile characters shall be located so that a clear floor area 18 inches (455 mm) minimum by 18 inches (455 mm) minimum, centered on the tactile characters, is provided beyond the arc of any door swing between the closed position and 45 degree open position.

EXCEPTION: Signs with tactile characters shall be permitted on the push side of doors with closers and without hold-open devices.

703.3.12 Finish and Contrast. Characters and their background shall have a nonglare finish. Characters shall contrast with their background with either light characters on a dark background, or dark characters on a light background.

EXCEPTION: Where separate tactile characters and visual characters with the same information are provided, tactile characters are not required to have nonglare finish or to contrast with their background.

703.4 Braille.

703.4.1 General. Braille shall be contracted (Grade 2) braille and shall comply with Section 703.4.

703.4.2 Uppercase Letters. The indication of an uppercase letter or letters shall only be used before the first word of sentences, proper nouns and names, individual letters of the alphabet, initials, or acronyms.

703.4.3 Dimensions. Braille dots shall have a domed or rounded shape and shall comply with Table 703.4.3.

703.4.4 Position. Braille shall be below the corresponding text. If text is multilined, braille shall be placed below entire text. Braille shall be separated $3/_8$ inch (9.5 mm) minimum from any other tactile characters and $3/_8$ inch (9.5 mm) minimum from raised borders and decorative elements. Braille provided on elevator car controls shall be separated $3/_{16}$ inch (4.8 mm) minimum either directly below or adjacent to the corresponding raised characters or symbols.

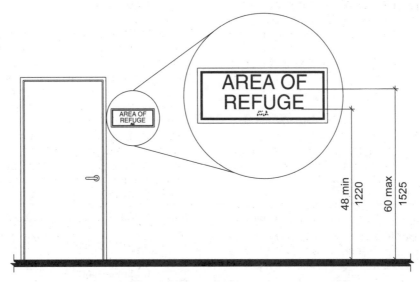

Note: For Braille mounting height see Section 703.4.5

Fig. 703.3.10
Height of Tactile Characters above Floor or Ground

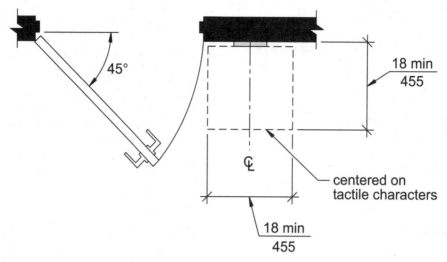

Fig. 703.3.11
Location of Tactile Signs at Doors

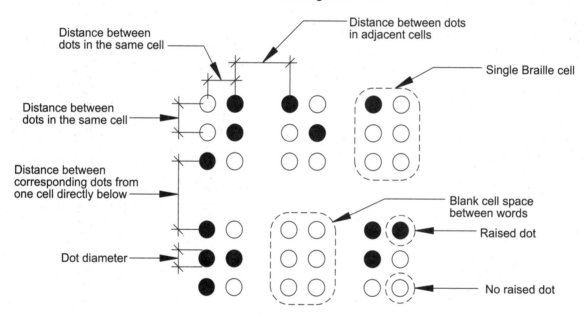

Fig. 703.4.3
Braille Measurement

Table 703.4.3— Braille Dimensions

Measurement range	Minimum in inches Maximum in inches
Dot base diameter	0.059 (1.5 mm) to 0.063 (1.6 mm)
Distance between two dots in the same cell	0.090 (2.3 mm) to 0.100 (2.5 mm)
Distance between corresponding dots in adjacent cells[1]	0.241 (6.1 mm) to 0.300 (7.6 mm)
Dot height	0.025 (0.6 mm) to 0.037 (0.9 mm)
Distance between corresponding dots from one cell directly below[1]	0.395 (10.0 mm) to 0.400 (10.2 mm)

[1]Measured center to center

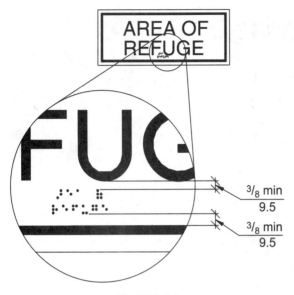

Fig. 703.4.4
Position of Braille

703.4.5 Mounting Height. Braille shall be 48 inches (1220 mm) minimum and 60 inches (1525 mm) maximum above the floor, measured to the baseline of the braille cells.

EXCEPTION: Elevator car controls shall not be required to comply with Section 703.4.5.

703.5 Pictograms.

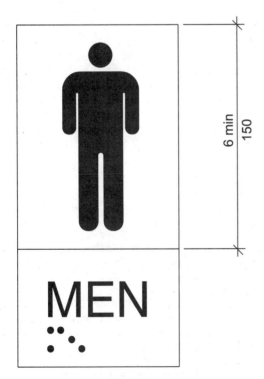

Fig. 703.5
Pictogram Field

703.5.1 General. Pictograms shall comply with Section 703.5.

703.5.2 Pictogram Field. Pictograms shall have a field 6 inches (150 mm) minimum in height. Characters or braille shall not be located in the pictogram field.

703.5.3 Finish and Contrast. Pictograms and their fields shall have a nonglare finish. Pictograms shall contrast with their fields, with either a light pictogram on a dark field or a dark pictogram on a light field.

703.5.4 Text Descriptors. Where text descriptors for pictograms are required, they shall be located directly below the pictogram field. Text descriptors shall comply with Sections 703.3 and 703.4.

703.6 Symbols of Accessibility.

703.6.1 General. Symbols of accessibility shall comply with Section 703.6.

703.6.2 Finish and Contrast. Symbols of accessibility and their backgrounds shall have a nonglare finish. Symbols of accessibility shall contrast with their backgrounds, with either a light symbol on a dark background or a dark symbol on a light background.

703.6.3 Symbols.

703.6.3.1 International Symbol of Accessibility. The International Symbol of Accessibility shall comply with Figure 703.6.3.1.

703.6.3.2 International Symbol of TTY. The International Symbol of TTY shall comply with Figure 703.6.3.2.

703.6.3.3 Assistive Listening Systems. Assistive listening systems shall be identified by the International Symbol of Access for Hearing Loss complying with Figure 703.6.3.3.

703.6.3.4 Volume-Controlled Telephones. Telephones with volume controls shall be identified by a pictogram of a telephone handset with radiating sound waves on a square field complying with Figure 703.6.3.4.

703.7 Remote Infrared Audible Sign (RIAS) Systems.

703.7.1 General. Remote Infrared Audible Sign Systems shall comply with Section 703.7.

703.7.2 Transmitters. Where provided, Remote Infrared Audible Sign Transmitters shall be designed to communicate with receivers complying with Section 703.7.3.

**Fig. 703.6.3.1
International Symbol of
Accessibility**

**Fig. 703.6.3.2
International TTY Symbol**

**Fig. 703.6.3.3
International Symbol of
Access for Hearing Loss**

**Fig. 703.6.3.4
Volume-Controlled
Telephone**

703.7.3 Remote Infrared Audible Sign Receivers.

703.7.3.1 Frequency. Basic speech messages shall be frequency modulated at 25 kHz, with a +/- 2.5 kHz deviation, and shall have an infrared wavelength from 850 to 950 nanometer (nm).

703.7.3.2 Optical Power Density. Receiver shall produce a 12 decibel (dB) signal-plus-noise-to-noise ratio with a 1 kHz modulation tone at +/- 2.5 kHz deviation of the 25 kHz subcarrier at an optical power density of 26 picowatts per square millimeter measured at the receiver photosensor aperture.

703.7.3.3 Audio Output. The audio output from an internal speaker shall be at 75 dBA minimum at 18 inches (455 mm) with a maximum distortion of 10 percent.

703.7.3.4 Reception Range. The receiver shall be designed for a high dynamic range and capable of operating in full-sun background illumination.

703.7.3.5 Multiple Signals. A receiver provided for the capture of the stronger of two signals in the receiver field of view shall provide a received power ratio on the order of 20 dB for negligible interference.

703.8 Pedestrian Signals. Accessible pedestrian signals shall comply with Section 4E.06 - Accessible Pedestrian Signals, and Section 4E.08 - Accessible Pedestrian Signal Detectors, of the Manual on Uniform Traffic Control Devices listed in Section 105.2.1.

EXCEPTION: Pedestrian signals are not required to comply with the requirement for choosing audible tones.

704 Telephones

704.1 General. Accessible public telephones shall comply with Section 704.

704.2 Wheelchair Accessible Telephones. Wheelchair accessible public telephones shall comply with Section 704.2.

704.2.1 Clear Floor Space. A clear floor space complying with Section 305 shall be provided. The clear floor space shall not be obstructed by bases, enclosures, or seats.

704.2.1.1 Parallel Approach. Where a parallel approach is provided, the distance from the edge of the telephone enclosure to the face of the telephone shall be 10 inches (255 mm) maximum.

704.2.1.2 Forward Approach. Where a forward approach is provided, the distance from the front edge of a counter within the enclosure to the face of the telephone shall be 20 inches (510 mm) maximum.

704.2.2 Operable Parts. The highest operable part of the telephone shall comply with Section 308. Telephones shall have push button controls where service for such equipment is available.

704.2.3 Telephone Directories. Where provided, telephone directories shall comply with Section 309.

704.2.4 Cord Length. The telephone handset cord shall be 29 inches (735 mm) minimum in length.

704.2.5 Hearing-Aid Compatibility. Telephones shall be hearing aid compatible.

704.3 Volume-Control Telephones. Public telephones required to have volume controls shall be equipped with a receive volume control that provides a gain adjustable up to 20 dB minimum. Incremental volume controls shall provide at least one intermediate step of gain of 12 dB minimum. An automatic reset shall be provided.

704.4 TTY. TTYs required at a public pay telephone shall be permanently affixed within, or adjacent to, the telephone enclosure. Where an acoustic coupler is used, the telephone cord shall be of sufficient length to allow connection of the TTY and the telephone handset.

704.5 Height. When in use, the touch surface of TTY keypads shall be 34 inches (865 mm) minimum above the floor.

EXCEPTION: Where seats are provided, TTYs shall not be required to comply with Section 704.5.

704.6 TTY Shelf. Where pay telephones designed to accommodate a portable TTY are provided, they shall be equipped with a shelf and an electrical outlet within or adjacent to the telephone enclosure. The telephone handset shall be capable of being placed flush on the surface of the shelf. The shelf shall be capable of accommodating a TTY and shall have a vertical clearance 6 inches (150 mm) minimum in height above the area where the TTY is placed.

704.7 Protruding Objects. Telephones, enclosures, and related equipment shall comply with Section 307.

705 Detectable Warnings

705.1 General. Detectable warning surfaces shall comply with Section 705.

705.2 Standardization. Detectable warning surfaces shall be standard within a building, facility, site, or complex of buildings.

EXCEPTION: In facilities that have both interior and exterior locations, detectable warnings in exterior locations shall not be required to comply with Section 705.4.

705.3 Contrast. Detectable warning surfaces shall contrast visually with adjacent surfaces, either light-on-dark or dark-on-light.

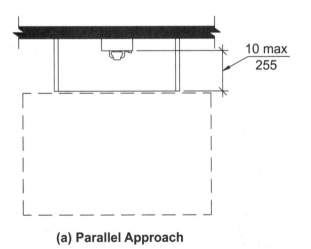

(a) Parallel Approach

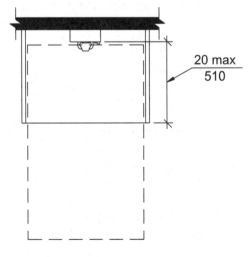

(b) Forward Approach

Fig. 704.2.1
Clear Floor Space for Telephones

705.4 Interior Locations. Detectable warning surfaces in interior locations shall differ from adjoining walking surfaces in resiliency or sound-on-cane contact.

705.5 Truncated Domes. Detectable warning surfaces shall have truncated domes complying with Section 705.5.

705.5.1 Size. Truncated domes shall have a base diameter of 0.9 inch (23 mm) minimum to 1.4 inch (36 mm) maximum, and a top diameter of 50 percent minimum to 65 percent maximum of the base diameter.

705.5.2 Height. Truncated domes shall have a height of 0.2 inch (5.1 mm).

705.5.3 Spacing. Truncated domes shall have a center-to-center spacing of 1.6 inches (41 mm) minimum and 2.4 inches (61 mm) maximum, and a base-to-base spacing of 0.65 inch (16.5 mm) minimum, measured between the most adjacent domes on the grid.

705.5.4 Alignment. Truncated domes shall be aligned in a square grid pattern.

706 Assistive Listening Systems

706.1 General. Accessible assistive listening systems in assembly areas shall comply with Section 706.

706.2 Receiver Jacks. Receivers required for use with an assistive listening system shall include a $^1/_8$-inch (3.2 mm) standard mono jack.

706.3 Receiver Hearing-Aid Compatibility. Receivers required to be hearing aid compatible shall interface with telecoils in hearing aids through the provision of neck loops.

706.4 Sound Pressure Level. Assistive listening systems shall be capable of providing a sound pressure level of 110 dB minimum and 118 dB maximum, with a dynamic range on the volume control of 50 dB.

706.5 Signal-to-Noise Ratio. The signal-to-noise ratio for internally generated noise in assistive listening systems shall be 18 dB minimum.

706.6 Peak Clipping Level. Peak clipping shall not exceed 18 dB of clipping relative to the peaks of speech.

707 Automatic Teller Machines (ATMs) and Fare Machines

707.1 General. Accessible automatic teller machines and fare machines shall comply with Section 707.

707.2 Clear Floor Space. A clear floor space complying with Section 305 shall be provided in front of the machine.

EXCEPTION: Clear floor space is not required at drive up only automatic teller machines and fare machines.

707.3 Operable Parts. Operable parts shall comply with Section 309. Each operable part shall be able to be differentiated by sound or touch, without activation.

EXCEPTION: Drive up only automatic teller machines and fare machines shall not be required to comply with Section 309.2 or 309.3.

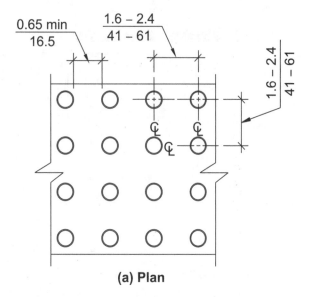

(a) Plan

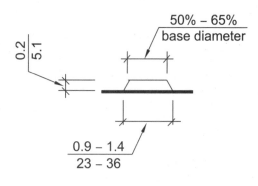

(b) Elevation (Enlarged)

Fig. 705.5
Truncated Dome Size and Spacing

707.4 Privacy. Automatic teller machines shall provide the opportunity for the same degree of privacy of input and output available to all individuals.

707.5 Numeric Keys. Numeric keys shall be arranged in a 12-key ascending or descending telephone keypad layout. The number Five key shall have a single raised dot.

707.6 Function Keys. Function keys shall comply with Section 707.6.

    707.6.1 Tactile Symbols. Function key surfaces shall have raised tactile symbols as shown in Table 707.6.1.

    707.6.2 Contrast. Function keys shall contrast visually from background surfaces. Characters and symbols on key surfaces shall contrast visually from key surfaces. Visual contrast shall be either light-on-dark or dark-on-light.

        EXCEPTION: Tactile symbols required by Section 707.6.1 shall not be required to comply with Section 707.6.2.

707.7 Display Screen. The display screen shall comply with Section 707.7.

    707.7.1 Visibility. The display screen shall be visible from a point located 40 inches (1015 mm) above the center of the clear floor space in front of the machine.

        EXCEPTION: Drive up only automatic teller machines and fare machines shall not be required to comply with Section 707.7.1.

Table 707.6.1—Tactile Symbols

Key Function	Description of Tactile Symbol	Tactile Symbol
Enter or Proceed:	CIRCLE	○
Clear or Correct:	LEFT ARROW	←
Cancel:	"X"	×
Add Value:	PLUS SIGN	+
Decrease Value:	MINUS SIGN	-

    707.7.2 Characters. Characters displayed on the screen shall be in a sans serif font. The uppercase letter "I" shall be used to determine the allowable height of all characters of the font. The uppercase letter "I" of the font shall be $3/16$ inch (4.8 mm) minimum in height. Characters shall contrast with their background with either light characters on a dark background, or dark characters on a light background.

707.8 Speech Output. Machines shall be speech enabled. Operating instructions and orientation, visible transaction prompts, user input verification, error messages, and all displayed information for full use shall be accessible to and independently usable by individuals with vision impairments. Speech shall be delivered through a mechanism that is readily available to all users including, but not limited to, an industry standard connector or a tele-

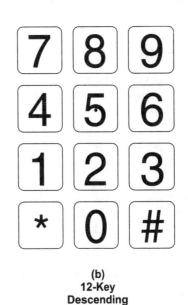

(a)
**12-Key
Ascending**

(b)
**12-Key
Descending**

**Fig. 707.5
Numeric Key Layout**

phone handset. Speech shall be recorded or digitized human, or synthesized.

EXCEPTIONS:

1. Audible tones shall be permitted in lieu of speech for visible output that is not displayed for security purposes, including but not limited to, asterisks representing personal identification numbers.

2. Advertisements and other similar information shall not be required to be audible unless they convey information that can be used in the transaction being conducted.

3. Where speech synthesis cannot be supported, dynamic alphabetic output shall not be required to be audible.

707.8.1 User Control. Speech shall be capable of being repeated and interrupted by the user. There shall be a volume control for the speech function.

EXCEPTION: Speech output for any single function shall be permitted to be automatically interrupted when a transaction is selected.

707.8.2 Receipts. Where receipts are provided, speech output devices shall provide audible balance inquiry information, error messages, and all other information on the printed receipt necessary to complete or verify the transaction.

EXCEPTIONS:

1. Machine location, date and time of transaction, customer account number, and the machine identifier shall not be required to be audible.

2. Information on printed receipts that duplicates audible information available on-screen shall not be required to be presented in the form of an audible receipt.

3. Printed copies of bank statements and checks shall not be required to be audible.

707.9 Input Controls. At least one tactually discernible input control shall be provided for each function. Where provided, key surfaces not on active areas of display screens shall be raised above surrounding surfaces. Where membrane keys are the only method of input, each shall be tactually discernable from surrounding surfaces and adjacent keys.

707.10 Braille Instructions. Braille instructions for initiating the speech mode shall be provided. Braille shall comply with Section 703.4.

708 Two-Way Communication Systems

708.1 General. Accessible two-way communication systems shall comply with Section 708.

708.2 Audible and Visual Indicators. The system shall provide both visual and audible signals.

708.3 Handsets. Handset cords, if provided, shall be 29 inches (735 mm) minimum in length.

Chapter 8. Special Rooms and Spaces

801 General

801.1 Scope. Special rooms and spaces required to be accessible by the scoping provisions adopted by the administrative authority shall comply with the applicable provisions of Chapter 8.

802 Assembly Areas

802.1 General. Wheelchair spaces and wheel chair space locations in assembly areas with spectator seating shall comply with Section 802.

802.2 Floor Surfaces. The floor surface of wheelchair space locations shall have a slope not steeper than 1:48 and shall comply with Section 302.

802.3 Width. A single wheelchair space shall be 36 inches (915 mm) minimum in width. Where two adjacent wheelchair spaces are provided, each wheelchair space shall be 33 inches (840 mm) minimum in width.

802.4 Depth. Where a wheelchair space location can be entered from the front or rear, the wheelchair space shall be 48 inches (1220 mm) minimum in depth. Where a wheelchair space location can only be entered from the side, the wheelchair space shall be 60 inches (1525 mm) minimum in depth.

802.5 Approach. The wheelchair space location shall adjoin an accessible route. The accessible route shall not overlap the wheelchair space location.

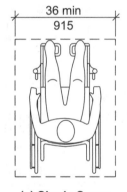

(a) Single Space

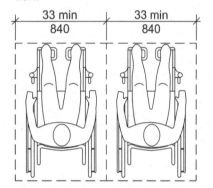

(b) Multiple Adjacent Spaces

Fig. 802.3
Width of a Wheelchair Space in Assembly Areas

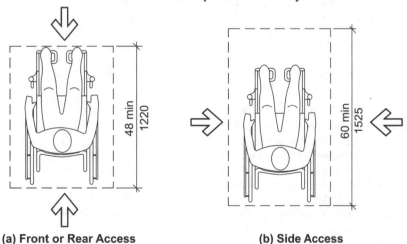

(a) Front or Rear Access   **(b) Side Access**

Fig. 802.4
Depth of a Wheelchair Space in Assembly Areas

802.5.1 Overlap. A wheelchair space location shall not overlap the required width of an aisle.

802.6 Integration of Wheelchair Space Locations. Wheelchair space locations shall be an integral part of any seating area.

802.7 Companion Seat. A companion seat, complying with Section 802.7, shall be provided beside each wheelchair space.

802.7.1 Companion Seat Type. The companion seat shall be comparable in size and quality to assure equivalent comfort to the seats within the seating area adjacent to the wheelchair space location. Companion seats shall be permitted to be moveable.

802.7.2 Companion Seat Alignment. In row seating, the companion seat shall be located to provide shoulder alignment with the wheelchair space occupant. The shoulder of the wheelchair space occupant is considered to be 36 inches (915 mm) from the front of the wheelchair space. The floor surface for the companion seat shall be at the same elevation as the wheelchair space floor surface.

802.8 Designated Aisle Seats. Designated aisle seats shall comply with Section 802.8.

802.8.1 Armrests. Where armrests are provided on seating in the immediate area of designated aisle seats, folding or retractable armrests shall be provided on the aisle side of the designated aisle seat.

802.8.2 Identification. Each designated aisle seat shall be identified by a sign or marker.

802.9 Lines of Sight. Where spectators are expected to remain seated for purposes of viewing events, spectators in wheelchair space locations shall be provided with a line of sight in accordance with Section 802.9.1. Where spectators in front of the wheelchair space locations will be expected to stand at their seats for purposes of viewing events, spectators in wheelchair space locations shall be provided with a line of sight in accordance with Section 802.9.2.

802.9.1 Line of Sight over Seated Spectators. Where spectators are expected to remain seated during events, spectators seated in wheelchair space locations shall be provided with lines of sight to the performance area or playing field comparable to that provided to spectators in closest proximity to the wheelchair space location. Where seating provides lines of sight over heads, spectators in wheelchair space locations shall be afforded lines of sight complying with Section

802.9.1.1. Where wheelchair space locations provide lines of sight over the shoulder and between heads, spectators in wheelchair space locations shall be afforded lines of sight complying with Section 802.9.1.2.

802.9.1.1 Lines of Sight over Heads. Spectators seated in wheelchair space locations shall be afforded lines of sight over the heads of seated individuals in the first row in front of the wheelchair space location.

802.9.1.2 Lines of Sight between Heads. Spectators seated in wheelchair space locations shall be afforded lines of sight over the shoulders and between the heads of seated individuals in the first row in front of the wheelchair space location.

802.9.2 Line of Sight over Standing Spectators. Wheelchair space locations required to provide a line of sight over standing spectators shall comply with Section 802.9.2.

802.9.2.1 Distance from Adjacent Seating. The front of the wheelchair space location shall be 12 inches (305 mm) maximum from the back of the chair or bench in front.

802.9.2.2 Elevation. The elevation of the tread on which a wheelchair space location is located shall comply with Table 802.9.2.2. For riser heights other than those provided, interpolations shall be permitted.

802.10 Wheelchair Space Dispersion. Wheelchair spaces shall be dispersed to the minimum number of locations in accordance with Table 802.10. Wheelchair space locations shall be dispersed in accordance with Sections 802.10.1, 802.10.2 and 802.10.3. In addition, in spaces utilized primarily for viewing motion picture projection, wheelchair space locations shall be dispersed in accordance with Section 802.10.4. Once the required number of wheelchair space locations has been met, further dispersion is not required.

802.10.1 Horizontal Dispersion. Wheelchair space locations shall be dispersed horizontally to provide viewing options. Locations shall be separated by a minimum of 10 intervening seats. Two wheelchair spaces shall be permitted to be located side-by-side.

EXCEPTION: In venues where wheelchair space locations are provided on only one side or on two opposite sides of the performance area or playing field, horizontal dispersion is not required where the locations are within the 2nd or 3rd quartile of the total row length. The wheelchair space locations and companion seats shall be permitted to overlap into the 1st and 4th quartile of the total row length if the

2nd and 3rd quartile of the row length does not provide the required length for the wheelchair space locations and companion seats. All intermediate aisles shall be included in determining the total row length.

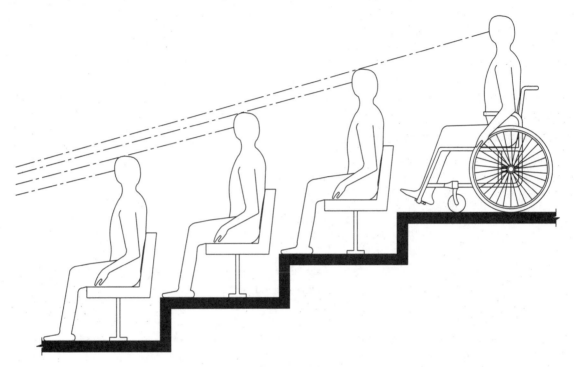

Fig. 802.9.1.1
Lines of Sight over the Heads of Seated Spectators

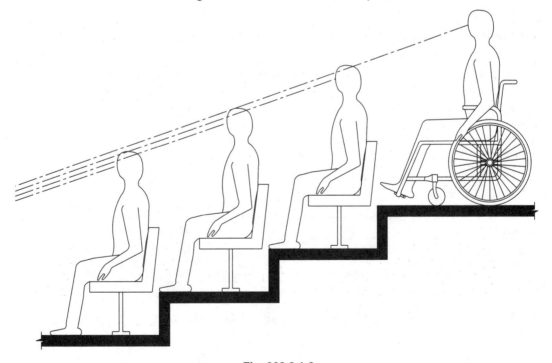

Fig. 802.9.1.2
Lines of Sight between the Heads of Seated Spectators

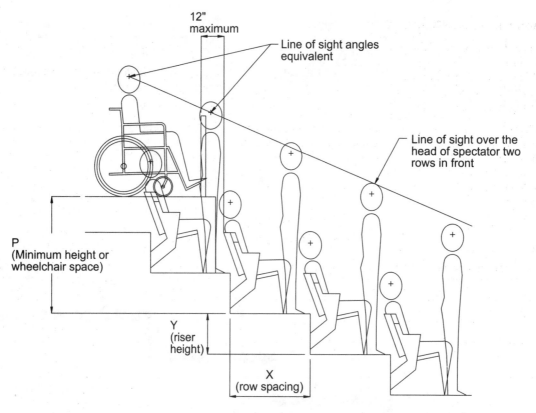

Calculation:

$$P = [(2X + 34)(Y - 2.25)/X] + (20.2 - Y)$$

Fig. 802.9.2
Wheelchair Space Elevation

Table 802.9.2.2
Required Wheelchair Space Location Elevation Over Standing Spectators

Riser height	Minimum height of the wheelchair space location based on row spacing[1]		
	Rows less than 33 inches (840 mm)[2]	Rows 33 inches (840 mm) to 44 inches (1120 mm)[2]	Rows over 44 inches (1120 mm)[2]
0 inch (0 mm)	16 inch (405 mm)	16 inch (405 mm)	16 inch (405 mm)
4 inch (102 mm)	22 inch (560 mm)	21 inch (535 mm)	21 inch (535 mm)
8 inch (205 mm)	31 inch (785 mm)	30 inch (760 mm)	28 inch (710 mm)
12 inch (305 mm)	40 inch (1015 mm)	37 inch (940 mm)	35 inch (890 mm)
16 inch (406 mm)	49 inch (1245 mm)	45 inch (1145 mm)	42 inch (1065 mm)
20 inch (510 mm)[3]	58 inch (1475 mm)	53 inch (1345 mm)	49 inch (1245 mm)
24 inch (610 mm)	N/A	61 inch (1550 mm)	56 inch (1420 mm)
28 inch (710 mm)[4]	N/A	69 inch (1750 mm)	63 inch (1600 mm)
32 inch (815 mm)	N/A	N/A	70 inch (1780 mm)
36 inch (915 mm) and higher	N/A	N/A	77 inch (1955 mm)

(continued)

Footnotes to Table 802.9.2.2

[1]The height of the wheelchair space location is the vertical distance from the tread of the row of seats directly in front of the wheelchair space location to the tread of the wheelchair space location.

[2]The row spacing is the back-to-back horizontal distance between the rows of seats in front of the wheelchair space location.

[3]Seating treads less than 33 inches (840 mm) in depth are not permitted with risers greater than 18 inches (455 mm) in height.

[4]Seating treads less than 44 inches (1120 mm) in depth are not permitted with risers greater than 27 inches (685 mm) in height.

NOTE: Table 802.8.9 is based on providing a spectator in a wheelchair a line of sight over the head of a spectator two rows in front of the wheelchair space location using average anthropometrical data. The table is based on the following calculation: $[(2X+34)(Y-2.25)/X]+(20.2-Y)$ where Y is the riser height of the rows in front of the wheelchair space location and X is the tread depth of the rows in front of the wheelchair space location. The calculation is based on the front of the wheelchair space location being located 12 inches (305 mm) from the back of the seating tread directly in front and the eye of the standing spectator being set back 8 inches (205 mm) from the riser.

Table 802.10
Wheelchair Space Dispersion

Total seating in Assembly Areas	Minimum required number of wheelchair spaces
Up to 150	1
151 to 500	2
501 to 1000	3
1001 to 5,000	3, plus 1 additional space for each 1,000 seats or portions thereof above 1,000
5,001 and over	7, plus 1 additional space for each 2,000 seats or portions thereof above 5,000

802.10.2 Dispersion for Variety of Distances from the Event. Wheelchair space locations shall be dispersed at a variety of distances from the event to provide viewing options. Locations shall be separated by a minimum of five intervening rows.

EXCEPTIONS:

1. In bleachers, wheelchair space locations shall not be required to be provided in rows other than rows at points of entry to bleacher seating.

2. In spaces utilized for viewing motion picture projections, assembly spaces with 300 seats or less shall not be required to comply with Section 802.10.2.

3. In spaces other than those utilized for viewing motion picture projections, assembly spaces with 300 seats or less shall not be required to comply with Section 802.10.2 if the wheelchair space locations are within the front 50 percent of the total rows.

802.10.3 Dispersion by Type. Where there are seating areas, each having distinct services or amenities, wheelchair space locations shall be provided within each seating area.

802.10.4 Spaces Utilized Primarily for Viewing Motion Picture Projections. In spaces utilized primarily for viewing motion picture projections, wheelchair space locations shall comply with Section 802.10.4.

802.10.4.1 Spaces with Seating on Risers. Where tiered seating is provided, wheelchair space locations shall be integrated into the tiered seating area.

802.10.4.2 Distance from the Screen. Wheelchair space locations shall be located within the rear 70 percent of the seats provided.

803 Dressing, Fitting, and Locker Rooms

803.1 General. Accessible dressing, fitting, and locker rooms shall comply with Section 803.

803.2 Turning Space. A turning space complying with Section 304 shall be provided within the room.

803.3 Door Swing. Doors shall not swing into the room unless a clear floor space complying with Section 305.3 is provided within the room, beyond the arc of the door swing.

803.4 Benches. A bench complying with Section 903 shall be provided within the room.

803.5 Coat Hooks and Shelves. Accessible coat hooks provided within the room shall accommodate a forward reach or side reach complying with Section 308. Where provided, a shelf shall be 40 inches (1015 mm) minimum and 48 inches (1220 mm) maximum above the floor.

804 Kitchens and Kitchenettes

804.1 General. Accessible kitchens and kitchenettes shall comply with Section 804.

804.2 Clearance. Where a pass-through kitchen is provided, clearances shall comply with Section 804.2.1. Where a U-shaped kitchen is provided, clearances shall comply with Section 804.2.2.

> **EXCEPTION:** Spaces that do not provide a cooktop or conventional range shall not be required to comply with Section 804.2.

804.2.1 Pass-through Kitchens. In pass-through kitchens where counters, appliances or cabinets are on two opposing sides, or where counters, appliances or cabinets are opposite a parallel wall, clearance between all opposing base cabinets, counter tops, appliances, or walls within kitchen work areas shall be 40 inches (1015 mm)

minimum. Pass-through kitchens shall have two entries.

804.2.2 U-Shaped Areas. In kitchens enclosed on three contiguous sides, clearance between all opposing base cabinets, countertops, appliances, or walls within kitchen work areas shall be 60 inches (1525 mm) minimum.

804.3 Work Surface. Work surfaces shall comply with Section 902.

> **EXCEPTION:** Spaces that do not provide a cooktop or conventional range shall not be required to provide an accessible work surface.

804.4 Sinks. Sinks shall comply with Section 606.

804.5 Storage. At least 50 percent of shelf space in cabinets shall comply with Section 905.

804.6 Appliances. Where provided, kitchen appliances shall comply with Section 804.6.

804.6.1 Clear Floor Space. A clear floor space complying with Section 305 shall be provided at each kitchen appliance. Clear floor spaces are permitted to overlap.

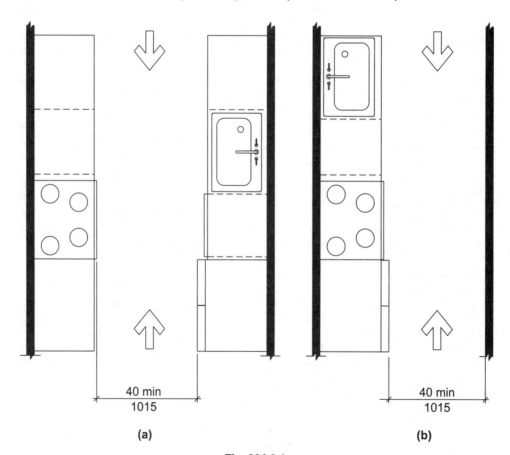

40 min
1015

(a)

40 min
1015

(b)

**Fig. 804.2.1
Pass-through Kitchen Clearance**

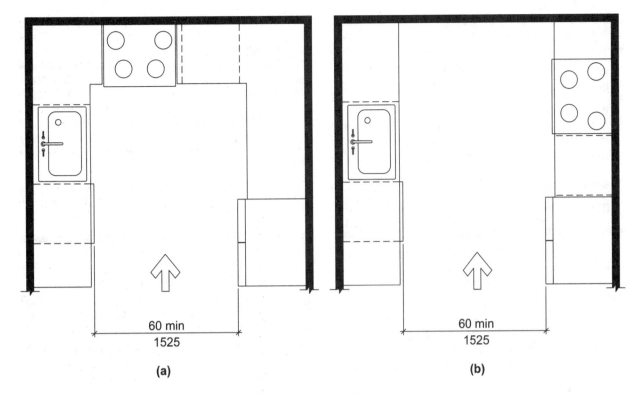

(a)

(b)

Fig. 804.2.2
U-Shaped Kitchen Clearance

804.6.2 Operable Parts. All appliance controls shall comply with Section 309.

EXCEPTIONS:

1. Appliance doors and door latching devices shall not be required to comply with Section 309.4.

2. Bottom-hinged appliance doors, when in the open position, shall not be required to comply with Section 309.3.

804.6.3 Dishwasher. A clear floor space, positioned adjacent to the dishwasher door, shall be provided. The dishwasher door in the open position shall not obstruct the clear floor space for the dishwasher or an adjacent sink.

804.6.4 Range or Cooktop. A clear floor space, positioned for a parallel or forward approach to the space for a range or cooktop, shall be provided. Where the clear floor space is positioned for a forward approach, knee and toe clearance complying with Section 306 shall be provided. Where knee and toe space is provided, the underside of the range or cooktop shall be insulated or otherwise configured to prevent burns, abrasions, or electrical shock. The location of controls shall not require reaching across burners.

804.6.5 Oven. Ovens shall comply with Section 804.6.5.

804.6.5.1 Side-Hinged Door Ovens. Side-hinged door ovens shall have a work surface complying with Section 804.3 positioned adjacent to the latch side of the oven door.

804.6.5.2 Bottom-Hinged Door Ovens. Bottom-hinged door ovens shall have a work surface complying with Section 804.3 positioned adjacent to one side of the door.

804.6.5.3 Controls. Ovens shall have controls on front panels.

804.6.6 Refrigerator/Freezer. Combination refrigerators and freezers shall have at least 50 percent of the freezer compartment shelves, including the bottom of the freezer, 54 inches (1370 mm) maximum above the floor when the shelves are installed at the maximum heights possible in the compartment. A clear floor space, positioned for a parallel approach to the space dedicated to a refrigerator/freezer, shall be provided. The centerline of the clear floor space shall be offset 24 inches (610 mm) maximum from the centerline of the dedicated space.

805 Transportation Facilities

805.1 General. Transportation facilities shall comply with Section 805.

805.2 Bus Boarding and Alighting Areas. Bus boarding and alighting areas shall comply with Section 805.2.

805.2.1 Surface. Bus stop boarding and alighting areas shall have a firm, stable surface.

805.2.2 Dimensions. Bus stop boarding and alighting areas shall have a 96 inches (2440 mm) minimum clear length, measured perpendicular to the curb or vehicle roadway edge, and a 60 inches (1525 mm) minimum clear width, measured parallel to the vehicle roadway.

805.2.3 Slope. The slope of the bus stop boarding and alighting area parallel to the vehicle roadway shall be the same as the roadway, to the maximum extent practicable. The slope of the bus stop boarding and alighting area perpendicular to the vehicle roadway shall be 1:48 maximum.

805.2.4 Connection. Bus stop boarding and alighting areas shall be connected to streets, sidewalks, or pedestrian paths by an accessible route complying with Section 402.

805.3 Bus Shelters. Bus shelters shall provide a minimum clear floor space complying with Section 305 entirely within the shelter. Bus shelters shall be connected by an accessible route complying with Section 402 to a boarding and alighting area complying with Section 805.2.

805.4 Bus Signs. Bus route identification signs shall have visual characters complying with Sections 703.2.2, 703.2.3, and 703.2.5 through 703.2.8. In addition, bus route identification numbers shall be visual characters complying with Section 703.2.4.

> **EXCEPTION:** Bus schedules, timetables and maps that are posted at the bus stop or bus bay shall not be required to comply with Section 805.4.

805.5 Rail Platforms. Rail platforms shall comply with Section 805.5.

805.5.1 Slope. Rail platforms shall not exceed a slope of 1:48 in all directions.

> **EXCEPTION:** Where platforms serve vehicles operating on existing track or track laid in existing roadway, the slope of the platform parallel to the track shall be permitted to be equal to the slope (grade) of the roadway or existing track.

805.5.2 Detectable Warnings. Platform boarding edges not protected by platform screens or guards shall have a detectable warning complying with Section 705, 24 inches (610 mm) in width, along the full length of the public use area of the platform.

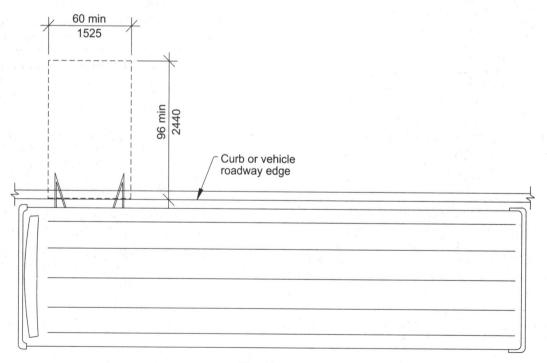

Fig. 805.2.2
Size of Bus Boarding and Alighting Areas

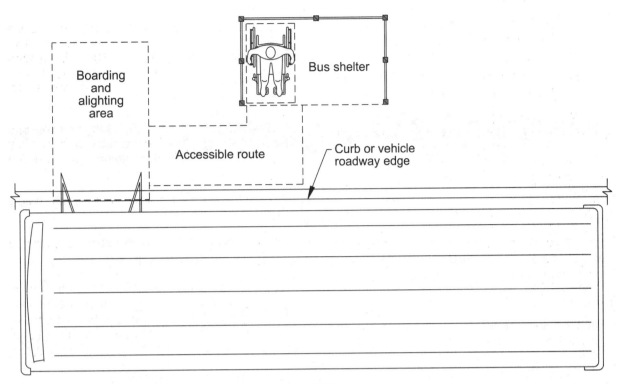

Fig. 805.3
Bus Shelters

805.6 Rail Station Signs. Rail station signs shall comply with Section 805.6.

EXCEPTION: Signs shall not be required to comply with Sections 805.6.1 and 805.6.2 where audible signs are remotely transmitted to hand-held receivers, or are user- or proximity-actuated.

805.6.1 Entrances. Where signs identify a station or a station entrance, at least one sign with tactile characters complying with Section 703.3 shall be provided at each entrance.

805.6.2 Routes and Destinations. Lists of stations, routes and destinations served by the station that are located on boarding areas, platforms, or mezzanines shall have visual characters complying with Section 703.2. A minimum of one tactile sign complying with Section 703.3 shall be provided on each platform or boarding area to identify the specific station.

EXCEPTION: Where sign space is limited, characters shall not be required to exceed 3 inches (76 mm) in height.

805.6.3 Station Names. Stations covered by this section shall have identification signs with visual characters complying with Section 703.2. The signs shall be clearly visible and within the sight lines of a standing or sitting passenger from within the vehicle on both sides when not obstructed by another vehicle.

805.7 Public Address Systems. Where public address systems convey audible information to the public, the same or equivalent information shall be provided in a visual format.

805.8 Clocks. Where clocks are provided for use by the public, the clock face shall be uncluttered so that its elements are clearly visible. Hands, numerals and digits shall contrast with the background either light-on-dark or dark-on-light. Where clocks are installed overhead, numerals and digits shall be visual characters complying with Section 703.2.

805.9 Escalators. Where provided, escalators shall have a 32-inch (815 mm) minimum clear width, and shall comply with Requirements 6.1.3.5.6 - Step Demarcations, and 6.1.3.6.5 - Flat Steps of ASME A17.1 listed in Section 105.2.5.

EXCEPTION: Existing escalators shall not be required to comply with Section 805.9.

805.10 Track Crossings. Where a circulation path serving boarding platforms crosses tracks, it shall comply with Section 402.

EXCEPTION: Openings for wheel flanges shall be permitted to be $2\frac{1}{2}$ inch (64 mm) maximum.

806 Holding Cells and Housing Cells

806.1 General. Holding cells and housing cells shall comply with Section 806.

Fig. 805.10
Track Crossings

806.2 Features for People Using Wheelchairs or Other Mobility Aids. Cells required to have features for people using wheelchairs or other mobility aids shall comply with Section 806.2.

806.2.1 Turning Space. Turning space complying with Section 304 shall be provided within the cell.

806.2.2 Benches. Where benches are provided, at least one bench shall comply with Section 903.

806.2.3 Beds. Where beds are provided, clear floor space complying with Section 305 shall be provided on at least one side of the bed. The clear floor space shall be positioned for parallel approach to the side of the bed.

806.2.4 Toilet and Bathing Facilities. Toilet facilities or bathing facilities provided as part of a cell shall comply with Section 603.

806.3 Communication Features. Cells required to have communication features shall comply with Section 806.3.

806.3.1 Alarms. Where audible emergency alarm systems are provided to serve the occupants of cells, visual alarms complying with Section 702 shall be provided.

EXCEPTION: In cells where inmates or detainees are not allowed independent means of egress, visual alarms shall not be required.

806.3.2 Telephones. Where provided, telephones within cells shall have volume controls complying with Section 704.3.

807 Courtrooms

807.1 General. Courtrooms shall comply with Section 807.

807.2 Turning Space. Where provided, each area that is raised or depressed and accessed by ramps or platform lifts with entry ramps shall provide an unobstructed turning space complying with Section 304.

807.3 Clear Floor Space. Within the defined area of each jury box and witness stand, a clear floor space complying with Section 305 shall be provided.

EXCEPTION: In alterations, wheelchair spaces are not required to be located within the defined area of raised jury boxes or witness stands and shall be permitted to be located outside these spaces where ramps or platform lifts restrict or project into the means of egress required by the administrative authority.

807.4 Judges' Benches and Courtroom Stations. Judges' benches, clerks' stations, bailiffs' stations, deputy clerks' stations, court reporters' stations and litigants' and counsel stations shall comply with Section 902.

Chapter 9. Built-In Furnishings and Equipment

901 General

901.1 Scope. Built-in furnishings and equipment required to be accessible by the scoping provisions adopted by the administrative authority shall comply with the applicable provisions of Chapter 9.

902 Dining Surfaces and Work Surfaces

902.1 General. Accessible dining surfaces and work surfaces shall comply with Section 902.

> **EXCEPTION:** Dining surfaces and work surfaces primarily for children's use shall be permitted to comply with Section 902.4.

902.2 Clear Floor Space. Clear floor space complying with Section 305, positioned for a forward approach, shall be provided. Knee and toe clearance complying with Section 306 shall be provided.

902.3 Height. The tops of dining surfaces and work surfaces shall be 28 inches (710 mm) minimum and 34 inches (865 mm) maximum in height above the floor.

902.4 Dining Surfaces and Work Surfaces for Children's Use. Accessible dining surfaces and work surfaces primarily for children's use shall comply with Section 902.4.

> **EXCEPTION:** Dining surfaces and work surfaces used primarily by children ages 5 and younger shall not be required to comply with Section 902.4 where a clear floor space complying with Section 305, positioned for a parallel approach, is provided.

902.4.1 Clear Floor Space. A clear floor space complying with Section 305, positioned for forward approach, shall be provided. Knee and toe clearance complying with Section 306 shall be provided.

> **EXCEPTION:** Knee clearance 24 inches (610 mm) minimum above the floor shall be permitted.

902.4.2 Height. The tops of tables and counters shall be 26 inches (660 mm) minimum and 30 inches (760 mm) maximum above the floor.

903 Benches

903.1 General. Accessible benches shall comply with Section 903.

903.2 Clear Floor Space. A clear floor space complying with Section 305, positioned for parallel approach to an end of the bench seat, shall be provided.

903.3 Size. Benches shall have seats 42 inches (1065 mm) minimum in length, and 20 inches (510 mm) minimum and 24 inches (610 mm) maximum in depth.

903.4 Back Support. The bench shall provide for back support or shall be affixed to a wall. Back support shall be 42 inches (1065 mm) minimum in length and shall extend from a point 2 inches (51 mm) maximum above the seat surface to a point 18 inches (455 mm) minimum above the seat surface. Back support shall be $2^1/_2$ inches (64 mm) maximum from the rear edge of the seat measured horizontally.

903.5 Height. The top of the bench seat shall be 17 inches (430 mm) minimum and 19 inches (485 mm) maximum above the floor, measured to the top of the seat.

903.6 Structural Strength. Allowable stresses shall not be exceeded for materials used where a vertical or horizontal force of 250 pounds (1112 N) is applied at any point on the seat, fastener mounting device, or supporting structure.

903.7 Wet Locations. Where provided in wet locations the surface of the seat shall be slip resistant and shall not accumulate water.

904 Sales and Service Counters

904.1 General. Accessible sales and service counters shall comply with Section 904 as applicable.

904.2 Approach. All portions of counters required to be accessible shall be located adjacent to a walking surface complying with Section 403.

904.3 Sales and Service Counters. Sales and service counters shall comply with Section 904.3.1 or 904.3.2. The accessible portion of the countertop shall extend the same depth as the sales and service countertop.

904.3.1 Parallel Approach. A portion of the counter surface 36 inches (915 mm) minimum in length and 36 inches (915 mm) maximum in height above the floor shall be provided. Where the counter surface is less than 36 inches (915 mm) in length, the entire counter surface shall be

36 inches (915 mm) maximum in height above the floor. A clear floor space complying with Section 305, positioned for a parallel approach adjacent to the accessible counter, shall be provided.

904.3.2 Forward Approach. A portion of the counter surface 30 inches (760 mm) minimum in length and 36 inches (915 mm) maximum in height above the floor shall be provided. A clear floor space complying with Section 305, positioned for a forward approach to the accessible counter, shall be provided. Knee and toe clearance complying with Section 306 shall be provided under the accessible counter.

904.4 Checkout Aisles. Checkout aisles shall comply with Section 904.4.

904.4.1 Aisle. Aisles shall comply with Section 403.

904.4.2 Counters. The checkout counter surface shall be 38 inches (965 mm) maximum in height above the floor. The top of the counter edge protection shall be 2 inches (51 mm) maximum above the top of the counter surface on the aisle side of the checkout counter.

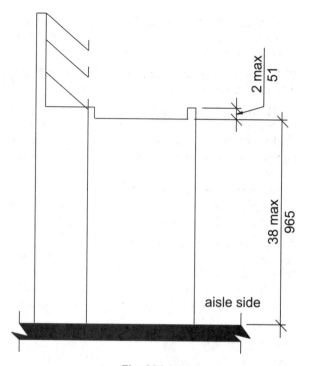

Fig. 904.4.2
Height of Checkout Counters

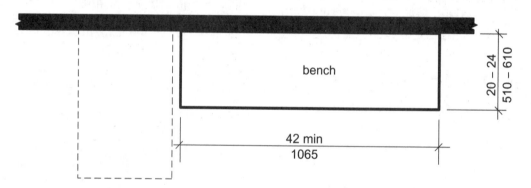

(a) Clear Floor Space and Size

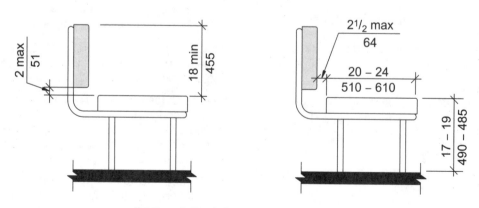

(b) Bench Back Support and Seat Height

Fig. 903
Benches

904.4.3 Check Writing Surfaces. Where provided, check writing surfaces shall comply with Section 902.3.

904.5 Food Service Lines. Counters in food service lines shall comply with Section 904.5.

904.5.1 Self-Service Shelves and Dispensing Devices. Self-service shelves and dispensing devices for tableware, dishware, condiments, food and beverages shall comply with Section 308.

904.5.2 Tray Slides. The tops of tray slides shall be 28 inches (710 mm) minimum and 34 inches (865 mm) maximum above the floor.

904.6 Security Glazing. Where counters or teller windows have security glazing to separate personnel from the public, a method to facilitate voice communication shall be provided. Telephone handset devices, if provided, shall comply with Section 704.3.

905 Storage Facilities

905.1 General. Accessible storage facilities shall comply with Section 905.

905.2 Clear Floor Space. A clear floor space complying with Section 305 shall be provided.

905.3 Height. Accessible storage elements shall comply with at least one of the reach ranges specified in Section 308.

905.4 Operable Parts. Operable parts of storage facilities shall comply with Section 309.

Chapter 10. Dwelling Units and Sleeping Units

1001 General

1001.1 Scoping. Dwelling units and sleeping units required to be Accessible units, Type A units, Type B units, or units with accessible communication features by the scoping provisions adopted by the administrative authority shall comply with the applicable provisions of Chapter 10.

1002 Accessible Units

1002.1 General. Accessible units shall comply with Section 1002.

1002.2 Primary Entrance. The accessible primary entrance shall be on an accessible route from public and common areas. The primary entrance shall not be to a bedroom.

1002.3 Accessible Route. Accessible routes within Accessible units shall comply with Section 1002.3. Exterior spaces less than 30 inches (760 mm) in depth or width shall comply with Sections 1002.3.1, 1002.3.3, 302, and 303.

> **1002.3.1 Location.** At least one accessible route shall connect all spaces and elements that are a part of the unit. Where only one accessible route is provided, it shall not pass through bathrooms and toilet rooms, closets, or similar spaces.
>
> > **EXCEPTION:** An accessible route is not required to unfinished attics and unfinished basements that are part of the unit.
>
> **1002.3.2 Turning Space.** All rooms served by an accessible route shall provide a turning space complying with Section 304.
>
> **1002.3.3 Components.** Accessible routes shall consist of one or more of the following elements: walking surfaces with a slope not steeper than 1:20, ramps, elevators, and platform lifts.

1002.4 Walking Surfaces. Walking surfaces that are part of an accessible route shall comply with Section 403.

1002.5 Doors and Doorways. The primary entrance door to the unit, and all other doorways intended for user passage, shall comply with Section 404.

> **EXCEPTION:** Existing doors to hospital patient sleeping rooms shall be exempt from the requirement for space at the latch side provided the door is 44 inches (1120 mm) minimum in width.

1002.6 Ramps. Ramps shall comply with Section 405.

1002.7 Elevators. Elevators within the unit shall comply with Section 407, 408, or 409.

1002.8 Platform Lifts. Platform lifts within the unit shall comply with Section 410.

1002.9 Operable Parts. Lighting controls, electrical switches and receptacle outlets, environmental controls, appliance controls, operating hardware for operable windows, plumbing fixture controls, and user controls for security or intercom systems shall comply with Section 309.

EXCEPTIONS:

1. Receptacle outlets serving a dedicated use.

2. One receptacle outlet shall not be required to comply with Section 309 where all of the following conditions are met:

   (a) the receptacle outlet is above a length of countertop that is uninterrupted by a sink or appliance;

   (b) at least one receptacle outlet complying with Section 1002.9 is provided for that length of countertop; and

   (c) all other receptacle outlets provided for that length of countertop comply with Section 1002.9.

3. Floor receptacle outlets.

4. HVAC diffusers.

5. Controls mounted on ceiling fans.

6. Where redundant controls other than light switches are provided for a single element, one control in each space shall not be required to be accessible.

1002.10 Laundry Equipment. Washing machines and clothes dryers shall comply with Section 611.

1002.11 Toilet and Bathing Facilities. Toilet and bathing facilities shall comply with Sections 603 through 610.

1002.11.1 Vanity Counter Top Space.
If vanity counter top space is provided in nonaccessible dwelling or sleeping units within the same facility, equivalent vanity counter top

space, in terms of size and proximity to the lavatory, shall also be provided in Accessible units.

1002.12 Kitchens. Kitchens shall comply with Section 804. At least one work surface, 30 inches (760 mm) minimum in length, shall comply with Section 902.

1002.13 Windows. Where operable windows are provided, at least one window in each sleeping, living, or dining space shall have operable parts complying with Section 1002.9. Each required operable window shall have operable parts complying with Section 1002.9.

1002.14 Storage Facilities. Where storage facilities are provided, they shall comply with Section 905. Kitchen cabinets shall comply with Section 804.5.

1003 Type A Units

1003.1 General. Type A units shall comply with Section 1003.

1003.2 Primary Entrance. The accessible primary entrance shall be on an accessible route from public and common areas. The primary entrance shall not be to a bedroom.

1003.3 Accessible Route. Accessible routes within Type A units shall comply with Section 1003.3. Exterior spaces less than 30 inches (760 mm) in depth or width shall comply with Sections 1003.3.1, 1003.3.3, 302, and 303.

1003.3.1 Location. At least one accessible route shall connect all spaces and elements that are a part of the unit. Where only one accessible route is provided, it shall not pass through bathrooms and toilet rooms, closets, or similar spaces.

EXCEPTION: An accessible route is not required to unfinished attics and unfinished basements that are part of the unit.

1003.3.2 Turning Space. All rooms served by an accessible route shall provide a turning space complying with Section 304.

EXCEPTION: Toilet rooms and bathrooms that are not required to comply with Section 1003.11.

1003.3.3 Components. Accessible routes shall consist of one or more of the following elements: walking surfaces with a slope not steeper than 1:20, ramps, elevators, and platform lifts.

1003.4 Walking Surfaces. Walking surfaces that are part of an accessible route shall comply with Section 403.

1003.5 Doors and Doorways. The primary entrance door to the unit, and all other doorways intended for user passage, shall comply with Section 404.

EXCEPTIONS:

1. Thresholds at exterior sliding doors shall be permitted to be $3/4$ inch (19 mm) maximum in height, provided they are beveled with a slope not greater than 1:2.

2. In toilet rooms and bathrooms not required to comply with Section 1003.11, maneuvering clearances required by Section 404.2.3 are not required on the toilet room or bathroom side of the door.

1003.6 Ramps. Ramps shall comply with Section 405.

1003.7 Elevators. Elevators within the unit shall comply with Section 407, 408, or 409.

1003.8 Platform Lifts. Platform lifts within the unit shall comply with Section 410.

1003.9 Operable Parts. Lighting controls, electrical switches and receptacle outlets, environmental controls, appliance controls, operating hardware for operable windows, plumbing fixture controls, and user controls for security or intercom systems shall comply with Section 309.

EXCEPTIONS:

1. Receptacle outlets serving a dedicated use.

2. One receptacle outlet is not required to comply with Section 309 where all of the following conditions are met:

    (a) the receptacle outlet is above a length of countertop that is uninterrupted by a sink or appliance; and

    (b) at least one receptacle outlet complying with Section 1003.9 is provided for that length of countertop; and

    (c) all other receptacle outlets provided for that length of countertop comply with Section 1003.9.

3. Floor receptacle outlets.

4. HVAC diffusers.

5. Controls mounted on ceiling fans.

6. Where redundant controls other than light switches are provided for a single element, one control in each space shall not be required to be accessible.

1003.10 Laundry Equipment. Washing machines and clothes dryers shall comply with Section 611.

1003.11 Toilet and Bathing Facilities.

1003.11.1 General. All toilet and bathing areas shall comply with Section 1003.11.9. At least one toilet and bathing facility shall comply with Section 1003.11. At least one lavatory, one water closet and either a bathtub or shower within the unit shall comply with Section 1003.11. The accessible toilet and bathing fixtures shall be in a single toilet/bathing area, such that travel between fixtures does not require travel through other parts of the unit.

1003.11.2 Doors. Doors shall not swing into the clear floor space or clearance for any fixture.

> **EXCEPTION:** Where a clear floor space complying with Section 305.3 is provided within the room beyond the arc of the door swing.

1003.11.3 Overlap. Clear floor spaces, clearances at fixtures and turning spaces are permitted to overlap.

1003.11.4 Reinforcement. Reinforcement shall be provided for the future installation of grab bars and shower seats at water closets, bathtubs, and shower compartments. Where walls are located to permit the installation of grab bars and seats complying with Sections 604.5, 607.4, 608.3 and 608.4, reinforcement shall be provided for the future installation of grab bars and seats meeting those requirements.

> **EXCEPTION:** Reinforcement is not required in a room containing only a lavatory and a water closet, provided the room does not contain the only lavatory or water closet on the accessible level of the dwelling unit.

1003.11.5 Lavatory. Lavatories shall comply with Section 606.

> **EXCEPTION:** Cabinetry shall be permitted under the lavatory, provided:
>
> (a) the cabinetry can be removed without removal or replacement of the lavatory;
>
> (b) the floor finish extends under such cabinetry; and
>
> (c) the walls behind and surrounding cabinetry are finished.

1003.11.6 Mirrors. Mirrors above lavatories shall have the bottom edge of the reflecting surface 40 inches (1015 mm) maximum above the floor.

1003.11.7 Water Closet. Water closets shall comply with Section 1003.11.7.

1003.11.7.1 Location. The water closet shall be positioned with a wall to the rear and to one side. The centerline of the water closet shall be 16 inches (405 mm) minimum and 18 inches (455 mm) maximum from the sidewall.

1003.11.7.2 Clearance. A clearance around the water closet of 60 inches (1525 mm) minimum, measured perpendicular from the side wall, and 56 inches (1420 mm) minimum, measured perpendicular from the rear wall, shall be provided.

1003.11.7.3 Overlap. The required clearance around the water closet shall be permitted to overlap the water closet, associated grab bars, paper dispensers, coat hooks, shelves, accessible routes, clear floor space required at other fixtures, and the wheelchair turning space. No other fixtures or obstructions shall be located within the required water closet clearance.

> **EXCEPTION:** A lavatory complying with Section 1003.11.5 shall be permitted on the rear wall 18 inches (455 mm) minimum from the centerline of the water closet where the clearance at the water closet is 66 inches (1675 mm) minimum measured perpendicular from the rear wall.

1003.11.7.4 Height. The top of the water closet seat shall be 15 inches (380 mm) minimum and 19 inches (485 mm) maximum above the floor, measured to the top of the seat.

1003.11.7.5 Flush Controls. Hand operated flush controls shall comply with Section 1003.9. Flush controls shall be located on the open side of the water closet.

1003.11.8 Bathtub. Bathtubs shall comply with Section 607.

> **EXCEPTIONS:**
>
> 1. The removable in-tub seat required by Section 607.3 is not required.
>
> 2. Counter tops and cabinetry shall be permitted at the control end of the clearance, provided such counter tops and cabinetry can be removed and the floor finish extends under such cabinetry.

1003.11.9 Shower. Showers shall comply with Section 608.

> **EXCEPTION:** Counter tops and cabinetry shall be permitted at the control end of the clearance, provided such counter tops and

cabinetry can be removed and the floor finish extends under such cabinetry.

1003.12 Kitchens. Kitchens shall comply with Section 1003.12.

1003.12.1 Clearance. Clearance complying with Section 1003.12.1 shall be provided.

1003.12.1.1 Minimum Clearance. Clearance between all opposing base cabinets, counter tops, appliances, or walls within kitchen work areas shall be 40 inches (1015 mm) minimum.

1003.12.1.2 U-Shaped Kitchens. In kitchens with counters, appliances, or cabinets on three contiguous sides, clearance between all opposing base cabinets, countertops, appliances, or walls within kitchen work areas shall be 60 inches (1525 mm) minimum.

1003.12.2 Clear Floor Space. Clear floor spaces required by Sections 1003.12.3 through 1003.12.6 shall comply with Section 305.

1003.12.3 Work Surface. At least one section of counter shall provide a work surface 30 inches (760 mm) minimum in length complying with Section 1003.12.3.

1003.12.3.1 Clear Floor Space. A clear floor space, positioned for a forward approach to the work surface, shall be provided. Knee and toe clearance complying with Section 306 shall be provided. The clear floor space shall be centered on the work surface.

EXCEPTION: Cabinetry shall be permitted under the work surface, provided:

(a) the cabinetry can be removed without removal or replacement of the work surface,

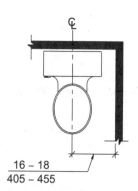

(a) Water Closet Location

16 – 18
405 – 455

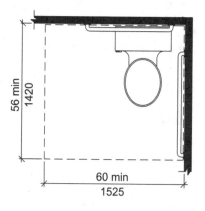

(b) Minimum Clearance

56 min
1420

60 min
1525

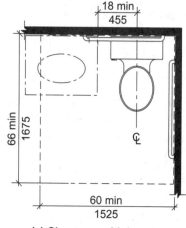

18 min
455

66 min
1675

60 min
1525

(c) Clearance with Lavatory
(Overlap Exception)

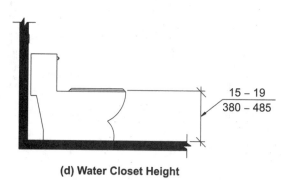

15 – 19
380 – 485

(d) Water Closet Height

Fig. 1003.11.7
Water Closets in Type A Units

(b) the floor finish extends under such cabinetry, and

(c) the walls behind and surrounding cabinetry are finished.

1003.12.3.2 Height. The work surface shall be 34 inches (865 mm) maximum above the floor.

EXCEPTION: A counter that is adjustable to provide a work surface at variable heights 29 inches (735 mm) minimum and 36 inches (915 mm) maximum above the floor, or that can be relocated within that range without cutting the counter or damaging adjacent cabinets, walls, doors, and structural elements, shall be permitted.

1003.12.3.3 Exposed Surfaces. There shall be no sharp or abrasive surfaces under the exposed portions of work surface counters.

1003.12.4 Sink. Sinks shall comply with Section 1003.12.4.

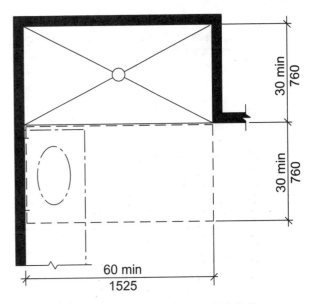

Note: Sink permitted per Section 608.2.2

Fig. 1003.11.9
Standard Roll-in-Type Shower
Compartment in Type A Units

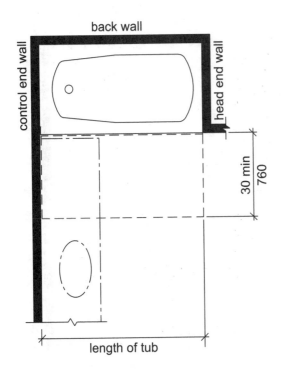

(a) Without Permanent Seat

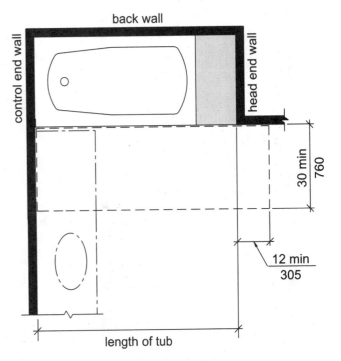

(b) With Permanent Seat

Fig. 1003.11.8
Clearance for Bathtubs in Type A Units

1003.12.4.1 Clear Floor Space. A clear floor space, positioned for a forward approach to the sink, shall be provided. Knee and toe clearance complying with Section 306 shall be provided. The clear floor space shall be centered on the sink bowl.

EXCEPTIONS:

1. The requirement for knee and toe clearance shall not apply to more than one bowl of a multi-bowl sink.

2. Cabinetry shall be permitted to be added under the sink, provided:

   (a) the cabinetry can be removed without removal or replacement of the sink,

   (b) the floor finish extends under such cabinetry, and

   (c) the walls behind and surrounding cabinetry are finished.

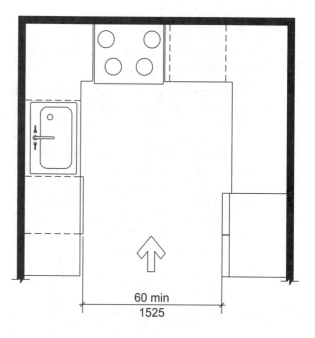

Fig. 1003.12.1.2
U-Shaped Kitchen Clearance in Type A Units

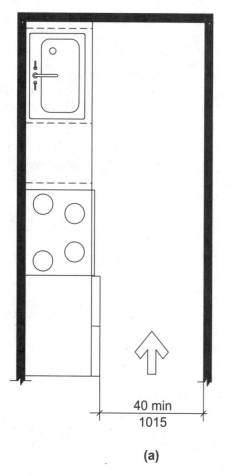

(a)

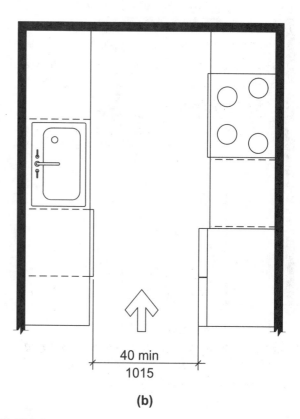

(b)

Fig. 1003.12.1.1
Minimum Kitchen Clearance in Type A Units

1003.12.4.2 Height. The front of the sink shall be 34 inches (865 mm) maximum above the floor, measured to the higher of the rim or counter surface.

EXCEPTION: A sink and counter that is adjustable to variable heights 29 inches (735 mm) minimum and 36 inches (915 mm) maximum above the floor, or that can be relocated within that range without cutting the counter or damaging adjacent cabinets, walls, doors and structural elements, provided rough-in plumbing permits connections of supply and drain pipes for sinks mounted at the height of 29 inches (735 mm), shall be permitted.

1003.12.4.3 Faucets. Faucets shall comply with Section 309.

1003.12.4.4 Exposed Pipes and Surfaces. Water supply and drain pipes under sinks shall be insulated or otherwise configured to protect against contact. There shall be no sharp or abrasive surfaces under sinks.

1003.12.5 Kitchen Storage. A clear floor space, positioned for a parallel or forward approach to the kitchen cabinets, shall be provided.

1003.12.6 Appliances. Where provided, kitchen appliances shall comply with Section 1003.12.6.

1003.12.6.1 Operable Parts. All appliance controls shall comply with Section 1003.9.

EXCEPTIONS:

1. Appliance doors and door latching devices shall not be required to comply with Section 309.4.

2. Bottom-hinged appliance doors, when in the open position, shall not be required to comply with Section 309.3.

1003.12.6.2 Clear Floor Space. A clear floor space, positioned for a parallel or forward approach, shall be provided at each kitchen appliance. Clear floor spaces shall be permitted to overlap.

1003.12.6.3 Dishwasher. A clear floor space, positioned adjacent to the dishwasher door, shall be provided. The dishwasher door in the open position shall not obstruct the clear floor space for the dishwasher or an adjacent sink.

1003.12.6.4 Range or Cooktop. A clear floor space, positioned for a parallel or forward approach to the space for a range or cooktop, shall be provided. Where the clear floor space is positioned for a forward approach, knee and toe clearance complying with Section 306 shall be provided. Where knee and toe space is provided, the underside of the range or cooktop shall be insulated or otherwise configured to protect from burns, abrasions, or electrical shock. The location of controls shall not require reaching across burners.

1003.12.6.5 Oven. Ovens shall comply with Section 1003.12.6.5. Ovens shall have controls on front panels, on either side of the door.

1003.12.6.5.1 Side-Hinged Door Ovens. Side-hinged door ovens shall have a countertop positioned adjacent to the latch side of the oven door.

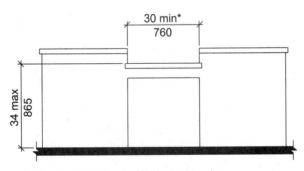

* 36 min. (915) if part of T-shaped turning space per Sections 304.3.2 and 1003.3.2

Fig. 1003.12.3
Work Surface in Kitchen for Type A Units

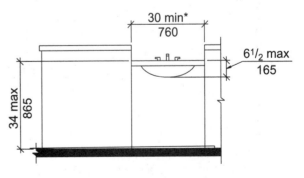

* 36 min. (915) if part of T-shaped turning space per Sections 304.3.2 and 1003.3.2

Fig. 1003.12.4
Kitchen Sink for Type A Units

1003.12.6.5.2 Bottom-Hinged Door Ovens. Bottom-hinged door ovens shall have a countertop positioned adjacent to one side of the door.

1003.12.6.6 Refrigerator/Freezer. Combination refrigerators and freezers shall have at least 50 percent of the freezer compartment shelves, including the bottom of the freezer 54 inches (1370 mm) maximum above the floor when the shelves are installed at the maximum heights possible in the compartment. A clear floor space, positioned for a parallel approach to the space dedicated to a refrigerator/freezer, shall be provided. The centerline of the clear floor space shall be offset 24 inches (610 mm) maximum from the centerline of the dedicated space.

1003.12.6.7 Trash Compactor. A clear floor space, positioned for a parallel or forward approach to the trash compactor, shall be provided.

1003.13 Windows. Where operable windows are provided, at least one window in each sleeping, living, or dining space shall have operable parts complying with Section 1003.9. Each required operable window shall have operable parts complying with Section 1003.9.

1003.14 Storage Facilities. Where storage facilities are provided, they shall comply with Section 1003.14. Kitchen cabinets shall comply with Section 1003.12.5.

1003.14.1 Clear Floor Space. A clear floor space complying with Section 305, positioned for a parallel or forward approach, shall be provided at each storage facility.

1003.14.2 Height. A portion of the storage area of each storage facility shall comply with at least one of the reach ranges specified in Section 308.

1003.14.3 Operable Parts. Operable parts on storage facilities shall comply with Section 309.

1004 Type B Units

1004.1 General. Type B units shall comply with Section 1004.

1004.2 Primary Entrance. The accessible primary entrance shall be on an accessible route from public and common areas. The primary entrance shall not be to a bedroom.

1004.3 Accessible Route. Accessible routes within Type B units shall comply with Section 1004.3.

1004.3.1 Location. At least one accessible route shall connect all spaces and elements that are a part of the unit. Where only one accessible route is provided, it shall not pass through bathrooms and toilet rooms, closets, or similar spaces.

EXCEPTION: One of the following is not required to be on an accessible route:

1. A raised floor area in a portion of a living, dining, or sleeping room; or

2. A sunken floor area in a portion of a living, dining, or sleeping room; or

3. A mezzanine that does not have plumbing fixtures or an enclosed habitable space.

1004.3.2 Components. Accessible routes shall consist of one or more of the following elements: walking surfaces with a slope not steeper than 1:20, doorways, ramps, elevators, and platform lifts.

1004.4 Walking Surfaces. Walking surfaces that are part of an accessible route shall comply with Section 1004.4.

1004.4.1 Width. Clear width of an accessible route shall comply with Section 403.5.

1004.4.2 Changes in Level. Changes in level shall comply with Section 303.

EXCEPTION: Where exterior deck, patio or balcony surface materials are impervious, the finished exterior impervious surface shall be 4 inches (100 mm) maximum below the floor level of the adjacent interior spaces of the unit.

1004.5 Doors and Doorways. Doors and doorways shall comply with Section 1004.5.

1004.5.1 Primary Entrance Door. The primary entrance door to the unit shall comply with Section 404.

EXCEPTION: Maneuvering clearances required by Section 404.2.3 shall not be required on the unit side of the primary entrance door.

1004.5.2 User Passage Doorways. Doorways intended for user passage shall comply with Section 1004.5.2.

1004.5.2.1 Clear Width. Doorways shall have a clear opening of $31^3/_4$ inches (810 mm) minimum. Clear opening of swinging doors shall be measured between the face of the door and stop, with the door open 90 degrees.

1004.5.2.2 Thresholds. Thresholds shall comply with Section 303.

> **EXCEPTION:** Thresholds at exterior sliding doors shall be permitted to be $3/_4$ inch (19 mm) maximum in height, provided they are beveled with a slope not steeper than 1:2.

1004.5.2.3 Automatic Doors. Automatic doors shall comply with Section 404.3.

1004.5.2.4 Double Leaf Doorways. Where an inactive leaf with operable parts higher than 48 inches (1220 mm) or lower than 15 inches (380 mm) above the floor is provided, the active leaf shall provide the clearance required by Section 1004.5.2.1.

1004.6 Ramps. Ramps shall comply with Section 405.

1004.7 Elevators. Elevators within the unit shall comply with Section 407, 408, or 409.

1004.8 Platform Lifts. Platform lifts within the unit shall comply with Section 410.

1004.9 Operable Parts. Lighting controls, electrical switches and receptacle outlets, environmental controls, and user controls for security or intercom systems shall comply with Sections 309.2 and 309.3.

EXCEPTIONS:

1. Receptacle outlets serving a dedicated use.

2. One receptacle outlet is not required to comply with Sections 309.2 and 309.3 where all of the following conditions are met:

   (a) the receptacle outlet is above a length of countertop that is uninterrupted by a sink or appliance; and

   (b) at least one receptacle outlet complying with Section 1004.9 is provided for that length of countertop; and

   (c) all other receptacle outlets provided for that length of countertop comply with Section 1004.9.

3. Floor receptacle outlets.

4. HVAC diffusers.

5. Controls mounted on ceiling fans.

6. Controls or switches mounted on appliances.

7. Plumbing fixture controls.

1004.10 Laundry Equipment. Washing machines and clothes dryers shall comply with Section 1004.10.

1004.10.1 Clear Floor Space. A clear floor space complying with Section 305.3, positioned for parallel approach, shall be provided. The clear floor space shall be centered on the appliance.

1004.11 Toilet and Bathing Facilities. Toilet and bathing fixtures shall comply with Section 1004.11.

> **EXCEPTION:** Fixtures on levels not required to be accessible.

1004.11.1 Clear Floor Space. Clear floor space required by Section 1004.11.3.1 or 1004.11.3.2 shall comply with Sections 1004.11.1 and 305.3.

1004.11.1.1 Doors. Doors shall not swing into the clear floor space for any fixture.

> **EXCEPTION:** Where a clear floor space complying with Section 305.3, excluding knee and toe clearances under elements, is provided within the room beyond the arc of the door swing.

1004.11.1.2 Knee and Toe Clearance. Clear floor space at fixtures shall be permitted to include knee and toe clearances complying with Section 306.

1004.11.1.3 Overlap. Clear floor spaces shall be permitted to overlap.

1004.11.2 Reinforcement. Reinforcement shall be provided for the future installation of grab bars and shower seats at water closets, bathtubs, and shower compartments. Where walls are located to permit the installation of grab bars and seats complying with Sections 604.5, 607.4, 608.3 and 608.4, reinforcement shall be provided for the future installation of grab bars and seats meeting those requirements.

> **EXCEPTION:** Reinforcement is not required in a room containing only a lavatory and a water closet, provided the room does not contain the only lavatory or water closet on the accessible level of the unit.

1004.11.3 Toilet and Bathing Rooms. Either all toilet and bathing rooms provided shall comply with Section 1004.11.3.1 (Option A), or one toilet and bathing room shall comply with Section 1004.11.3.2 (Option B).

1004.11.3.1 Option A. Each fixture provided shall comply with Section 1004.11.3.1.

> **EXCEPTION:** A lavatory and a water closet in a room containing only a lavatory and

water closet, provided the room does not contain the only lavatory or water closet on the accessible level of the unit.

1004.11.3.1.1 Lavatory. A clear floor space complying with Section 305.3, positioned for a parallel approach, shall be provided. The clear floor space shall be centered on the lavatory.

EXCEPTIONS:

1. A lavatory complying with Section 606.

2. Cabinetry shall be permitted under the lavatory provided such cabinetry can be removed without removal or replacement of the lavatory, and the floor finish extends under such cabinetry.

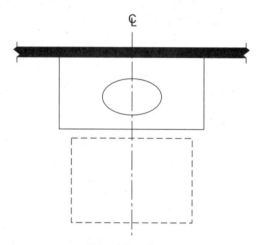

Clear Floor Space

**Fig. 1004.11.3.1.1
Lavatory in Type B Units—
Option A Bathrooms**

1004.11.3.1.2 Water Closet. The lateral distance from the centerline of the water closet to a bathtub or lavatory shall be 18 inches (455 mm) minimum on the side opposite the direction of approach and 15 inches (380 mm) minimum on the other side. The lateral distance from the centerline of the water closet to an adjacent wall shall be 18 inches (455 mm). The lateral distance from the centerline of the water closet to a lavatory or bathtub shall be 15 inches (380 mm) minimum. The water closet shall be positioned to allow for future installation of a grab bar on the side with 18 inches (455 mm) clearance. Clearance

around the water closet shall comply with Section 1004.11.3.1.2.1, 1004.11.3.1.2.2, or 1004.11.3.1.2.3.

1004.11.3.1.2.1 Parallel Approach. A clearance 56 inches (1420 mm) minimum measured from the wall behind the water closet, and 48 inches (1220 mm) minimum measured from a point 18 inches (455 mm) from the centerline of the water closet on the side designated for future installation of grab bars shall be provided. Vanities or lavatories on the wall behind the water closet are permitted to overlap the clearance.

1004.11.3.1.2.2 Forward Approach. A clearance 66 inches (1675 mm) minimum measured from the wall behind the water closet, and 48 inches (1220 mm) minimum measured from a point 18 inches (455 mm) from the centerline of the water closet on the side designated for future installation of grab bars shall be provided. Vanities or lavatories on the wall behind the water closet are permitted to overlap the clearance.

1004.11.3.1.2.3 Parallel or Forward Approach. A clearance 56 inches (1420 mm) minimum measured from the wall behind the water closet, and 42 inches (1065 mm) minimum measured from the centerline of the water closet shall be provided.

1004.11.3.1.3 Bathing Facilities. Where a bathtub or shower compartment is provided it shall conform with Section 1004.11.3.1.3.1, 1004.11.3.1.3.2, or 1004.11.3.1.3.3.

1004.11.3.1.3.1 Parallel Approach Bathtubs. A clearance 60 inches (1525 mm) minimum in length and 30 inches (760 mm) minimum in width shall be provided in front of bathtubs with a parallel approach. Lavatories complying with Section 606 shall be permitted in the clearance. A lavatory complying with Section 1004.11.3.1.1 shall be permitted at the control end of the bathtub if a clearance 48 inches (1220 mm) minimum in length and 30 inches (760 mm) minimum in width for a parallel approach is provided in front of the bathtub.

1004.11.3.1.3.2 Forward Approach Bathtubs. A clearance 60 inches (1525 mm) minimum in length and 48 inches (1220 mm) minimum in width shall be provided in front of bathtubs with a forward approach. A water closet shall be permitted in the clearance at the control end of the bathtub.

1004.11.3.1.3.3 Shower Compartment. If a shower compartment is the only bathing facility, the shower compartment shall have dimensions of 36 inches (915 mm) minimum in width and 36 inches (915 mm) minimum in depth. A clearance of 48 inches (1220 mm) minimum in length, measured perpendicular from the shower head wall, and 30 inches (760 mm) minimum in depth, measured from the face of the shower

compartment, shall be provided. Reinforcing for a shower seat is not required in shower compartments larger than 36 inches (915 mm) in width and 36 inches (915 mm) in depth.

1004.11.3.2 Option B. One of each type of fixture provided shall comply with Section 1004.11.3.2. The accessible fixtures shall be in a single toilet/bathing area, such that travel between fixtures does not require travel through other parts of the unit.

1004.11.3.2.1 Lavatory. Lavatories shall comply with Section 1004.11.3.2.1.

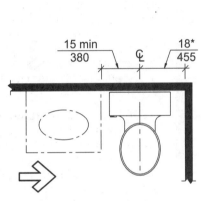

*18 min (455) to a fixture

(a) Water Closet Location

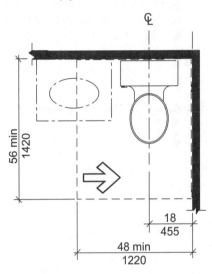

(b) Parallel Approach

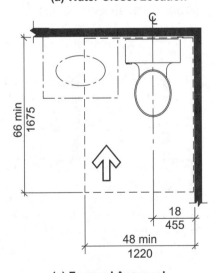

(c) Forward Approach

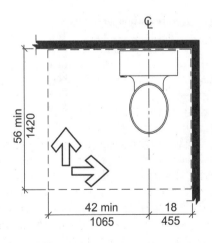

(d) Parallel or Forward Approach

**Fig. 1004.11.3.1.2
Water Closets in Type B Units**

1004.11.3.2.1.1 Clear Floor Space. A clear floor space complying with Section 305.3, positioned for a parallel approach, shall be provided.

EXCEPTIONS:

1. A lavatory complying with Section 606.

2. Cabinetry shall be permitted under the lavatory, provided such cabinetry can be removed without removal or replacement of the lavatory, and the floor finish extends under such cabinetry.

1004.11.3.2.1.2 Position. The clear floor space shall be centered on the lavatory.

1004.11.3.2.1.3 Height. The front of the lavatory shall be 34 inches (865 mm) maximum above the floor, measured to the higher of the fixture rim or counter surface.

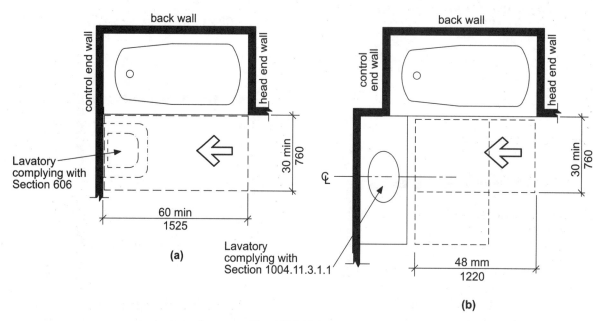

Fig. 1004.11.3.1.3.1
Parallel Approach Bathtub in Type B Units—Option A Bathrooms

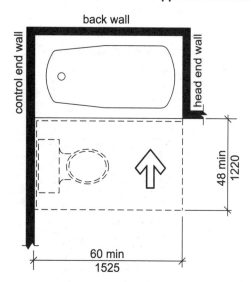

Fig. 1004.11.3.1.3.2
**Forward Approach Bathtub in Type B Units—
Option A Bathrooms**

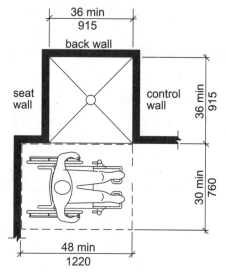

Fig. 1004.11.3.1.3.3
**Transfer-Type Shower Compartment in
Type B Units**

1004.11.3.2.2 Water Closet. The water closet shall comply with Section 1004.11.3.1.2.

1004.11.3.2.3 Bathing Facilities. Where either a bathtub or shower compartment is provided, it shall conform with Section 1004.11.3.2.3.1 or 1004.11.3.2.3.2.

1004.11.3.2.3.1 Bathtub. A clearance 48 inches (1220 mm) minimum in length measured perpendicular from the control end of the bathtub, and 30 inches (760 mm) minimum in width shall be provided in front of bathtubs.

1004.11.3.2.3.2 Shower Compartment. A shower compartment shall comply with Section 1004.11.3.1.3.3.

1004.12 Kitchens. Kitchens shall comply with Section 1004.12.

1004.12.1 Clearance. Clearance complying with Section 1004.12.1 shall be provided.

1004.12.1.1 Minimum Clearance. Clearance between all opposing base cabinets, counter tops, appliances, or walls within kitchen work areas shall be 40 inches (1015 mm) minimum.

1004.12.1.2 U-Shaped Kitchens. In kitchens with counters, appliances, or cabinets on three contiguous sides, clearance between all opposing base cabinets, countertops, appliances, or walls within kitchen work areas shall be 60 inches (1525 mm) minimum.

1004.12.2 Clear Floor Space. Clear floor space at appliances shall comply with Sections 1004.12.2 and 305.3.

1004.12.2.1 Sink. A clear floor space, positioned for a parallel approach to the sink, shall be provided. The clear floor space shall be centered on the sink bowl.

EXCEPTION: Sinks complying with Section 606 shall be permitted to have a clear floor space positioned for a parallel or forward approach.

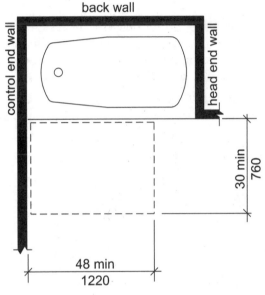

Fig. 1004.11.3.2.3.1
Bathroom Clearance in Type B Units—
Option B Bathrooms

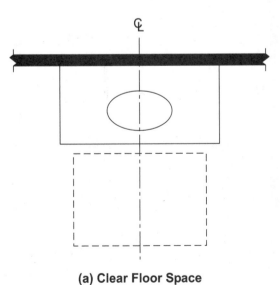

(a) Clear Floor Space

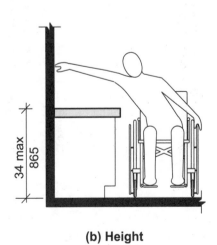

(b) Height

Fig. 1004.11.3.2.1
Lavatory in Type B Units—Option B Bathrooms

1004.12.2.2 Dishwasher. A clear floor space, positioned for a parallel or forward approach to the dishwasher, shall be provided. The clear floor space shall be positioned beyond the swing of the dishwasher door.

1004.12.2.3 Cooktop. A clear floor space, positioned for a parallel or forward approach to the cooktop, shall be provided. The centerline of the clear floor space shall align with the centerline of the cooktop. Where the clear floor space is positioned for a forward approach, knee and toe clearance complying with Section 306 shall be provided. Where knee and toe space is provided, the underside of the range or cooktop shall be insulated or otherwise configured to prevent burns, abrasions, or electrical shock.

1004.12.2.4 Oven. A clear floor space, positioned for a parallel or forward approach to the oven, shall be provided.

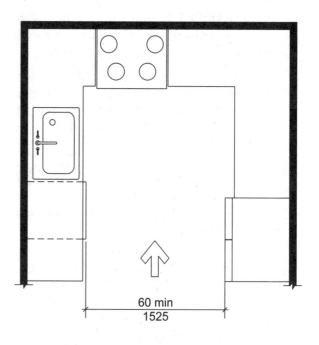

Fig. 1004.12.1.2
U-Shaped Kitchen Clearance in Type B Units

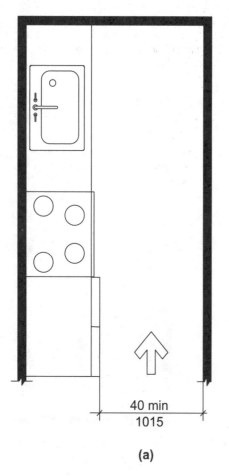

(a)

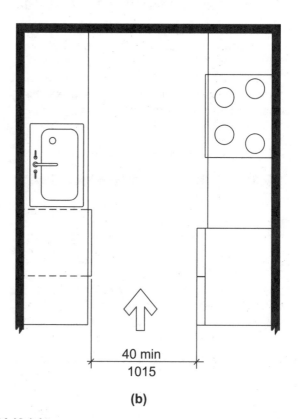

(b)

Fig. 1004.12.1.1
Minimum Kitchen Clearance in Type B Units

1004.12.2.5 Refrigerator/Freezer. A clear floor space, positioned for a parallel or forward approach to the refrigerator/freezer, shall be provided.

1004.12.2.6 Trash Compactor. A clear floor space, positioned for a parallel or forward approach to the trash compactor, shall be provided.

1005 Units with Accessible Communication Features

1005.1 General. Units required to have accessible communication features shall comply with Section 1005.

1005.2 Unit Smoke Detection. Where provided, unit smoke detection shall include audible notification complying with NFPA 72 listed in Section 105.2.2.

1005.3 Building Fire Alarm System. Where a building fire alarm system is provided, the system wiring shall be extended to a point within the unit in the vicinity of the unit smoke detection system.

1005.4 Visible Notification Appliances. Visible notification appliances, where provided within the unit as part of the unit smoke detection system or the building fire alarm system, shall comply with Section 1005.4.

1005.4.1 Appliance. Visible notification appliances shall comply with Section 702.

1005.4.2 Activation. All visible notification appliances provided within the unit for smoke detection notification shall be activated upon smoke detection. All visible notification appliances provided within the unit for building fire alarm notification shall be activated upon activation of the building fire alarm in the portion of the building containing the unit.

1005.4.3 Interconnection. The same visible notification appliances shall be permitted to provide notification of unit smoke detection and building fire alarm activation.

1005.4.4 Prohibited Use. Visible notification appliances used to indicate unit smoke detection or building fire alarm activation shall not be used for any other purpose within the unit.

1005.5 Unit Primary Entrance. Communication features shall be provided at the unit primary entrance complying with Section 1005.5.

1005.5.1 Notification. A hard-wired electric doorbell shall be provided. A button or switch shall be provided on the public side of the unit primary entrance. Activation of the button or switch shall initiate an audible tone within the unit.

1005.5.2 Identification. A means for visually identifying a visitor without opening the unit entry door shall be provided. Peepholes, where used, shall provide a minimum 180-degree range of view.

1005.6 Site, Building, or Floor Entrance. Where a system permitting voice communication between a visitor and the occupant of the unit is provided at a location other than the unit entry door, the system shall comply with Section 1005.6.

1005.6.1 Public or Common-Use Interface. The public or common-use system interface shall include the capability of supporting voice and TTY communication with the unit interface.

1005.6.2 Unit Interface. The unit system interface shall include a telephone jack capable of supporting voice and TTY communication with the public or common-use system interface.

1005.7 Closed-Circuit Communication Systems. Where a closed-circuit communication system is provided, the public or common-use system interface shall comply with Section 1005.6.1, and the unit system interface in units required to have accessible communication features shall comply with Section 1005.6.2.